choosing and keeping chickens

choosing and keeping chickens

Chris Graham

Printed and bound in China
06 07 08 09 10 1 3 5 7 9 8 6 4 2

Library of Congress Cataloging-in-Publication Data
Graham, Chris, 1961-
Choosing and keeping chickens / Chris Graham.
p. cm.
ISBN 0-7938-0601-1 (alk. paper)
1. Chickens. I. Title.
SF487.G6366 2006
636.5–dc22
2006023888

The Leader in Responsible Animal Care
for Over 50 Years!™
www.tfh.com

Contents

Introduction

The fact that you are reading this book probably means that you have decided that you want to own some chickens. This, in turn, means that your life is about to change for the better.

First steps

Chicken's demand attention, but they repay their owners with a range of benefits. They will cheer you up, fascinate and charm you, and, once you have gotten your first few birds, you will wonder why on earth you didn't start to keep them sooner.

Above *Keeping chickens will enthral you and your family. They make a wonderful addition to the household.*

It would be wrong to give the impression that it will all be clear sailing, however, because things can and do go wrong, particularly for beginners. In many ways, chickens are hardy creatures—much tougher than their appearance might suggest—but they are sensitive, too. Their environment plays a crucial part in their well-being, and the responsibility for maintaining it correctly lies entirely with their keeper. Get it wrong—by overcrowding the birds, supplying inadequate or inappropriate food and insufficient water, or allowing parasites to gain a hold, for example—and they will quickly become stressed. Unfortunately, the more stressed a chicken is, the more susceptible he becomes to other problems, such as disease and infections, and chickens can go downhill surprisingly quickly when mistakes are made.

However, if you follow the basic principles of good poultry care and management, you cannot go wrong. After all, poultry keeping is not exactly rocket science. Certainly, experience can count for a lot, but everyone has to start somewhere. Some old hands will have you believe that there are many secrets to be learned and that success will come only after years of painstaking work, but the reality is a good deal less daunting. Although breeding the perfect show specimen obviously requires both dedication and knowledge, it is perfectly possible for a complete beginner to live with, and enjoy, a small flock of productive chickens in the backyard with the minimum of fuss. And that, in essence, is what this book is all about, so read on and learn how to really enjoy your birds.

Left *Caring for chickens is not difficult and can be enormously rewarding.*

In the beginning...

Our relationship with the chicken can be traced back thousands of years, but unfortunately, it seems that our early interest was not especially beneficial for the birds, for many of the earliest records link chickens to fighting, notably among the ancient Greeks and the Romans. The ancient Egyptians kept fowl, too, but the birds that all these early civilizations knew were quite different from what we commonly think of as chickens. A good number of the popular breeds that are with us today are comparatively recent creations, with many appearing only in the past 150 years. Interest was spurred in Britain when poultry showing started to become fashionable in Victorian times. The "farmyard" fowl of the period were used as breeding stock and mixed with anything interesting that could be brought in from abroad.

There is one thing that all these breeds have in common, however, and that is the fact that they share the same ancestral roots. Most experts agree that the wild Jungle Fowl from Southeast Asia is the ancestor of today's domesticated breeds. The native species still survive today in reasonable numbers, living naturally in the forests of Indonesia. They are also kept in captivity, so their survival seems assured.

Jungle Fowl were domesticated by the people of the Indus Valley in India as early as 3000 BC. However, it took a long time for the news to spread, and the species didn't arrive in Egypt until the 15th century BC. Then, after a further 600 years or so, domesticated fowl reached the Mediterranean countries and were carried throughout western Europe during the Roman conquests.

There remain four distinct breeds of Jungle Fowl, the most significant of which is the Red Jungle Fowl (*Gallus bankiva*), which is believed to have had the single greatest influence on chicken domestication. By today's standards, this species is a pretty plain breed, with the male bird having essentially black and red plumage, although with some lighter coloring on the hackles (the feathers of the neck) and on the back. The female is even plainer, being mostly brown with a lighter breast.

Chickens in the west

How the Jungle Fowl was transformed into the 100 or so popular breeds we have today is a long and involved story, much of it lost in the mists of time. But the fact that it was is something for which we should all be grateful!

Farmyard fowl

To get an appreciation of the Jungle Fowl, you need look no further than the Old English Game, which, in appearance and temperament, is a reasonably similar bird. The Old English Game bird's ability as a fighter meant that the breed has been kept through the ages. Mainstream attitudes to cock fighting are different today, although it still goes on around the world, particularly in the Far East. It has become a largely covert activity elsewhere, even in those countries where it is legal.

For centuries in the West, chickens were simply kept as farmyard fowl. These birds were the descendants of those introduced by the Romans or possibly by the Vikings, and they were used as a casual source of eggs and the occasional bird for the pot. They included breeds such as the Dorking and Hamburgh, but there was next to no transfer of birds from area to area at this time, so groups inbred and distinct breeds developed on a regional basis. Poultry keeping in those days was a fairly insular, utilitarian activity.

Above *A pair of beautiful Sebrights; a true bantam with no large fowl counterpart.*

Then, in the early to mid-1800s, when more exotic breeds, such as the Cochin, Malay, and Brahma, started trickling into Europe from the Far East, things began to change, and poultry keeping became a popular activity.

At first, these exotic creatures were the exclusive preserve of wealthy landowners, for whom the rarity and unusual plumage made them into status symbols. Crossbreeding inevitably started between the established European breeds and these fancy newcomers, and the process of creating a whole new range of poultry breeds began—breeds that are now known as the pure breeds.

Eggs galore

Breeders soon realized the potential of the large, egg-laying breeds from the Mediterranean region, such as the Leghorn, which originated in Italy. Crossing these birds with indigenous fowl from elsewhere in Europe and America, as well as with some of the Asiatic breeds, led to the creation of the "utility" classification—birds that could be relied on to produce large numbers of top-quality eggs. Notable examples include the Rhode Island Red, the Sussex, and the Welsummer.

Inevitably, big business became increasingly involved in poultry farming—after all, there was money to be made. The world's ever-increasing demand for cheaper eggs and chicken meat drove specialist breeders into progressively more involved and secretive breeding programs, and the result was a new generation of "laying machines," known as the hybrids. Much of this pioneering work was done in the United States in the 1940s and 1950s, and today chickens are raised on an industrial scale around the world.

Think first

Keeping chickens in your own backyard is more than just a hobby. People who start raising them find that it quickly becomes a way of life. Their lifestyles are enriched by these addictive creatures and by everything that affects their everyday welfare.

Above *The Buff Sussex is a fine, traditional breed and also a good layer.*

However, the popularity of poultry keeping is reaching such levels now that chickens are becoming a "must-have" garden accessory. Of course, this is great for the hobby as a whole, but it can create problems when people leap into the idea without enough planning. Preparation and forethought are boring, but, as far as chickens are concerned, good preparation is the key to success.

The best poultry keepers are those who take their responsibilities seriously. Chickens are not the most demanding creatures, but they do require a reasonable level of care and attention if they are going to live happy and productive lives. So take your time, read this book, research things on the Internet, talk to other keepers and breeders... and then make the decision to get your first birds. Your efforts won't be wasted.

Which Breeds?

After you have made up your mind that you want to keep chickens and are certain that you will be able to do whatever is necessary to keep them healthy and happy, your next job is to decide what you want to get, and with more than 100 breeds to choose from, you are sure to find a breed that's right for you.

Choosing your chickens

First-time keepers should, if they are realistic, immediately reject a large proportion of the named breeds that are available simply on the grounds of their unsuitability. Picking the right breed is more involved than you might think.

Eliminating the obvious

Some breeds are very rare and hard to find; others are difficult to manage for anyone apart from those with specialist experience. Then there are breeds that are known to be nervous and excitable—they are often termed "flighty"—and these birds are best avoided if you have a young family and, perhaps, other pets.

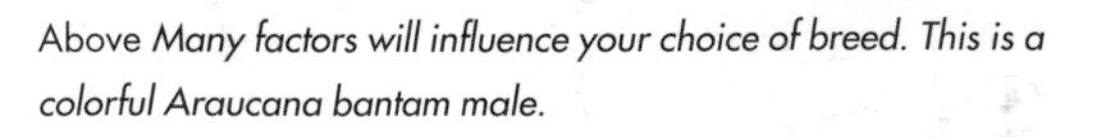

Above *Many factors will influence your choice of breed. This is a colorful Araucana bantam male.*

Nevertheless, even with these nonstarters crossed off your list, there are still plenty left to choose from. A good way of easing the decision-making process is to decide exactly what you want from your birds. Wanting a good supply of eggs plus the occasional bird for the table will make a big difference when deciding which breed you should buy. Alternatively, you might yearn to keep chickens for exhibition purposes. Feather color and patterning might be important to you as well. Size is another important consideration for many would-be owners, especially if the space you have available is limited.

New or old?

Divisions exist in the poultry-keeping world, with much depending on the type of birds you have and what you do with them. In many countries, a poultry establishment sets the standards to which breeds must conform. In the United States, the American Poultry Association is responsible for publishing these standards. In Britain, the Poultry Club of Great Britain oversees breed-based competitions and also appoints and approves the judges who adjudicate at prestigious shows and exhibitions.

If you are keen on the traditional approach or you are after exciting plumage combinations or eggs with unusual shell colors, you will want to consider the established, pure breeds—Araucana, Cochin, Orpington, Sussex, and Wyandotte, for example. Some of the pure breeds, such as the Dorking and the Indian Game, make tasty table birds, too.

Right *Indian Game are stunning birds, but are they right for you?*

Great layers

There are many practical factors that affect egg production, including age, health, and environment, so it is difficult to generalize about what to expect from individual birds. However, there are some guidelines that can be useful as a starting point.

Tradition or utility?
Any breed of fowl that originated in or around the Mediterranean region can usually be relied upon to deliver the goods. Leghorns, Anconas, and Minorcas, for example, will all normally be good layers. But these are by no means the only pure-breed options if you like a freshly boiled egg every morning for breakfast. The utility breeds can offer an even more prolific return, with the American Rhode Island Red, Wyandotte, and New Hampshire Red being particularly impressive. The same applies to the British Sussex and the Dutch Welsummer. These breeds were developed to provide the mainstay of the world's commercial egg production during the post-war era.

Some of the less productive pure breeds, such as the Cochin, have suffered at the hands of the exhibition fraternity. Over the years these chickens have been repeatedly bred for size, shape, and feathering above all else. A modern Cochin, with his giant featherduster-like plumage, bears little resemblance to the bird that

Above *Collecting fresh eggs is a satisfying advantage to keeping chickens. But egg production varies from breed to breed.*

arrived in Britain from the Far East during the 1840s. In fact, this was the breed that, almost single-handedly, sparked the exhibition movement in Britain. It was an instant success, particularly once word got out that Queen Victoria was an enthusiastic owner. The disadvantage, however, was that the Cochin lost his usefulness as a practical performer.

In fact, if your choice is going to be based on out-and-out laying performance, you need look no further than a hybrid chicken. These "laying machines" really have been designed for the job, and they will reward their owners with an almost limitless supply of eggs. These birds are not exhibited, and they still tend to divide poultry enthusiasts. The traditionalists regard them with a certain amount of disdain, but their growing following among backyard keepers is perfectly understandable. These birds are cheap to buy and are easy to look after. They are straightforward, no-nonsense birds, and they can be an ideal first choice.

Above *Bantam versions are just about a quarter of the size of their large fowl counterparts, as shown by these two Welsummer females.*

Little or large?

The popularity of bantams seems to grow hand-in-hand with the poultry-keeping hobby as a whole. Moreover, it is probably the case that a number of the clubs that cater to the rarer breeds have their bantam-owning members to thank for keeping them afloat.

Bantams are essentially small chickens, although there is a bit more to it than that. The group is subdivided into true bantams and other small versions, which are often referred to as miniatures. The distinction between the two is that true bantams have no large version, whereas the miniatures are scaled-down examples of the standard breeds. Officially, they should be no more than one-quarter the weight of the bird from which they are derived.

The name bantam is taken from the sea port in Java, from where the original examples of these tiny fowl were first exported about 200 years ago. Nowadays, virtually every one of the popular pure breeds has a bantam version. The true bantams are rather more limited and are characterized by their short backs, jaunty appearance, and wings, which point down toward the ground. Good examples of the type include the Sebright, the Japanese, and the Rosecomb.

Whether you choose a miniature or a true bantam, the great appeal of these birds is the combination of small size and practicality. If space is at a premium, bantams could be the answer. They can look every bit as spectacular as the larger fowl and will lay as well as their larger counterparts (although the eggs are obviously smaller). They are great family birds, too.

There are very few downsides to owning bantams. Egg size might be an issue for you, and if you are looking for a bird that will provide an occasional roast dinner, bantams are probably not the ideal starting point. In all other respects, these diminutive birds are an extremely sensible choice.

Meat or eggs?

Most people who start to keep chickens do so because they want a supply of delicious, healthy, fresh eggs. However, choosing one of the best-laying breeds is not the whole story.

Birds for eggs

Of the pure breeds, those that tend to make the best layers are the ones that show the least inclination to go "broody," which is when a hen settles to incubate a clutch of eggs. The instinct to do this is much stronger in some breeds than others, and those that are noted for doing it are termed "good sitters." The problem, as far as laying is concerned, is that during the time that the hen is broody, production ceases. This can mean several months without eggs from that bird.

The best sitting breeds tend to be the larger ones—Orpingtons and Sussex, for example—although there are plenty of exceptions to this rule. The Silkie, despite being one of the smallest breeds, is widely regarded as one of the best broodies of all. In general, the best laying performance comes from the smaller bodied, nonsitting breeds, especially those with a reputation for being active and hardy.

Environment plays a big part in how well hens lay, but keepers should appreciate that the influence of this extends as far as climate and even soil type. Some breeds, like the Minorca, do not perform as well in harsh climates, while others, like the Scots Dumpy, are more resilient to rougher weather and terrain.

Above *The Rhode Island Red is a good dual-purpose breed, known around the world for both its egg-laying and meat-producing qualities.*

Above *Modern hybrid breeds, like these Fenton Blues, have been developed to produce large numbers of eggs.*

Above *Good health (and egg-laying) depends on good-quality food and clean, fresh water, no matter what the breed of bird.*

Obviously, these factors will not apply to most backyards, but it is worth bearing in mind, nonetheless.

The laying performance of any hen changes with her age. All will produce the greatest number of eggs in their first year, when they are classified as pullets. They begin at about five or six months old (called the point-of-lay, POL). The first season's total is then reduced by about 20 percent annually in the second and third years, then falls away progressively after that. Egg production slows right down over the winter months for most birds, although the hybrids and the good pure-breed strains are better than most at continuing to lay.

The number of daylight hours is the problem for most birds, because egg-laying is stimulated by the amount of light and by day length. The shortening days heralding the onset of winter are sensed by receptors in the brain, and egg production slows and often stops. Some keepers reduce the effects of this by fitting lights inside their henhouses, which add an hour or two of light, morning and night. This can fool the birds into continuing to lay. Good-quality food (layers pellets or mash and some mixed grit) and a daily supply of fresh, clean water are also important factors.

Birds for the table

The idea of actually eating a chicken is one of the things that is usually furthest from the thoughts of most new keepers. Nevertheless, a proportion do rear for this purpose from the outset, while others become interested in the idea as they gain experience and start breeding their own birds. More and more owners are turning to their home-reared birds as a source of top-quality, additive- and chemical-free meat. The flavor to be enjoyed from a bird that has lived an unstressed, free-range life is far superior to the essentially tasteless supermarket product. The latter may be cheap, but that's just about all you can say for it.

If your aim is to produce table birds, you should be considering the "sitting" varieties—the big ones that go broody—and the Sussex is one of the best of these. Alternatively, if you take the Dorking and cross it with an Indian Game the result is a terrific table bird. These two old British breeds combine beautifully to produce tasty meat.

A more modern approach is to opt for one of the new breeds of hybrid "meat" birds, such as the French Sasso. They are fast to mature and will grow to a good size given the right environment and plenty of food.

It is important to get the feeding regime right though. Those keepers considering this option should take advice from good suppliers about feed rates, life spans, and so on. There are welfare issues to be considered, too.

Getting Started

Chickens thrive in good conditions, but, conversely, they are intolerant when things aren't right. If you want your birds to lay well and to live full and healthy lives, you must give them the right facilities to do so.

Space and shelter

Although keeping chickens is great fun, it has its serious side, too. Anyone who is thinking of getting involved with the hobby must appreciate that it comes hand-in-hand with a responsibility to the birds.

Committed to the cause

Never forget that the hens you keep will be entirely reliant on you for their good health and survival. They should never be bought on a whim, no matter how attractive and tempting the whole idea may appear. If you are in any doubt about the facilities you have or your own ability to provide the necessary care and attention, do not buy the birds in the first place.

Above *Your chickens will need to be fed and watered every day, without fail.*

The worst thing to do is to get everything established and then find that it is all too much to cope with and you don't want them after all.

Keeping chickens at home is not a hugely labor-intensive business, but it does require regular daily and watchful attention. The birds will need to be fed and watered every day. The fantasy image of hens pecking for corn at your feet on a sunny summer's day is not quite so idyllic on a cold, blustery winter morning. The henhouse will require regular cleaning to keep it habitable and to avoid infestations of pests.

Being a good poultry keeper requires you to develop a degree of stockmanship. It is like a sixth sense for the well-being of your flock, which enables you to recognize when things aren't right. Everyone can certainly learn the basics, but there is no substitute for experience.

It is a good idea to talk to other keepers first. Get an idea of the workload involved and whether you could work the daily routine required into your own lifestyle.

Housing your hens

The two most basic requirements for a healthy environment are space and shelter. Chickens need to roost inside at night, both for their own comfort (away from wind and rain) and to provide them with protection against natural predators. A secure, dry, and well-ventilated henhouse is essential. There is plenty of choice nowadays, with many manufacturers producing all sorts of different designs in all shapes and sizes. Some are traditional, others are modern. Prices vary enormously, too, as does quality, so it is important to make the right choice.

Above *Your birds will need adequate space and shelter to thive.*

The henhouse probably represents the biggest poultry-related investment you will make, which is why it is sensible to research the subject thoroughly. Be careful, however, because the information provided by manufacturers may not necessarily be in the best interest of your chickens. For a more realistic assessment, talk to other keepers or members of your local poultry society or club. Buying based on a recommendation that is founded on the experience of others is generally the best way. You will know that the information you have been given will be offered for all the right reasons and with no incentive to make a sale.

Location, location, location

Before you do anything else, you need to plan where you are going to site the henhouse. Think about how it is going to fit into your garden or backyard, and also spare a thought for your neighbors. Getting off on the wrong foot with the people next door can cause no end of trouble. It makes a lot of sense to discuss your plans with them and to get their approval right from the start. Remember, it takes only a couple of complaints from disgruntled neighbors to stir up real trouble. Also, bear in mind that a large henhouse—if it is the size of a garden shed, for example—may require a permit or the approval of your local authority.

Left *Henhouses must be sturdy and well made to ensure a dry but well-ventilated environment inside.*

Choosing a henhouse

The golden rule when buying a henhouse is to get the largest one that your budget and garden will allow. The overcrowding of chickens is something to be avoided at all costs.

The right enclosure

Chicken houses come in all shapes and sizes—simple arks (with an A-frame structure), pent-roofed (sloping), and apex-roofed. Some henhouses are even made of plastic, but traditionally wood is the building material of choice for the domestic keeper because it is durable (especially if made of tannalized wood), repairable, and relatively affordable.

Above *Henhouses can be simple or ornate. This one features an integral run and raised roosting compartment—ideal for these Serama bantams.*

The current popularity of poultry keeping means that many companies have established themselves as housing manufacturers. Inevitably, some of them peddle badly made products that should be avoided. The type and thickness of the wood used, the quality of the jointing, and the fit of the doors all provide useful pointers to a product's overall quality. Your henhouse is going to need to last, so reject any that appear flimsy and are of lightweight construction.

At its most basic, a poultry enclosure must guarantee its occupants shelter from the wind and rain, a dry floor, good amounts of natural light, and protection from draughts, while still allowing adequate ventilation. Inside, it should feature nest boxes in which the birds will lay their eggs. Sometimes these will be built on to the outside of the house to maximize interior space, but they should always be at the darker end of the roosting compartment and out of any direct sunlight.

Allow one nest box for every three or four birds. Ideally, each should be 12 inches (30 cm) square and 8–10 inches (20–25 cm) deep for large fowl and smaller for bantams. If they are bigger than this, there is a risk that more than one bird at a time might use them, and this will result in broken eggs. If they are not at floor level, it is a good idea if there is an alighting rail along the front edge of the nest boxes to help the hens get in and out. The position of the rail will depend on the height of the boxes above the floor, and thought must also be given to heavy breeds th might find it difficult to make the jump. If you keep these breeds, fitting a ladder or ramp will make all the difference to the ease with which they can enter and exit.

Above left *External access to the nest boxes is a great convenience factor for easy egg collecting.*
Above right *This henhouse has an external nest box and attached run.*

Room to move

In addition to the henhouse, the birds will require a run. A scratching shed, a covered run with a solid roof and sides, allows birds to scratch and exercise in dry, airy conditions, and they can be allowed out for supervised exercise in summer. A roosting shed and a scratching shed are the best options for backyards because they avoid muddy patches.

When space is very limited, the run can take the form of a netted wooden frame attached to the side of the henhouse; but for larger numbers of birds (more than six), you must build a proper enclosure. This needs to be securely fenced to keep the birds in and predators out. In some cases, the run may also benefit from some sort of roofing.

Much will depend on the type of chickens you intend to keep. No chickens can fly properly for any distance, but some can clear fences and hedges that are 6 feet (2 meters) high, and those that are known to be fliers, such as Anconas and Hamburghs, need to be contained with a netting roof. Birds with feathered legs and feet or ornate crests are best kept under a solid roof to keep the run area dry.

No chicken likes living in a wet, muddy environment, so do not site runs in low-lying, boggy places. Areas that may be lovely and grassy during the summer months can quickly turn to a quagmire once winter arrives, and this will be bad news for the birds. If your land is badly drained, you might have to construct a special winter run, with a solid roof and deep-litter floor covering (untreated softwood chips or dry sand are best for this purpose). The covering should be at least 6 inches (15 cm) deep, and you must not allow it to get damp or soiled.

Above *If you can't let your birds out, an integral run is the next best thing, assuming, of course, that you don't overcrowd it.*

Inside the henhouse

To keep your chickens happy inside the henhouse, you must provide them with somewhere to perch and with sufficient space to move around. Many enclosures are just too small for the number of hens inside them.

Perching potential

Make sure that there is at least one (preferably more) roosting perch on which the birds will sleep at night. The thickness of the perch is very important. Ideally, it should be at least 2 inches (5 cm) wide for large fowl and slightly narrower for smaller breeds. In addition, it should have rounded-off corners on its top edges. Perches that are too narrow cause the birds discomfort.

If more than one perch is fitted, make sure that they are all at the same height. Hens will naturally pick the highest roosting point, so offering a choice of perch height will end in an unnecessary and potentially damaging squabble. Make sure, too, that the perches are not too close to the side wall, otherwise the birds' tail feathers could be damaged.

Perches should be positioned well above floor level (and higher than the nest boxes), otherwise the birds will sleep in the nest boxes and make the eggs dirty. They should also be sited over a removable droppings tray. About half of a chicken's manure output occurs at night, so catching it on an extractable tray makes a lot of sense. Levels of ammonia (from the feces) will inevitably build up over night at floor level. The birds need to be kept above this, which is why adequate

Left *Most birds will roost on perches inside the enclosure, and these must be at least 2 inches (5 cm) wide for large fowl like these Sussex hens.*

Above *A removable droppings tray makes it easy to keep the roosting area clean.*

perch height and proper ventilation are important. Again, larger breeds can find the distance a struggle, so a ramp or ladder will be sensible. Heavy birds sometimes damage their feet and legs by repeatedly jumping down from a high perch.

Externally there will be what is called a pop hole, which provides the birds with a way in and out of the henhouse. Most modern designs feature a sliding cover for this that runs up and down in channels on either side so that the enclosure can be shut at night. There will also be access to the nest boxes from the outside to ease egg collection, plus a large door (or removable panel) so that you can get inside to clean and disinfect the henhouse.

A room with a view

Hens enjoy having a certain amount of natural light in their enclosures, and some of the "luxury" structures at the top end of the scale offer windows to provide this as well as ventilation. These aren't usually glazed but are covered with wire mesh and have an external wooden shutter that can be opened or closed depending on the weather. The mesh, of course, leaves the henhouse secure for night-time opening in summer and keeps out wild birds.

Ventilation is a key factor, especially if the number of birds living in the enclosure is running close to its design capacity, and it is particularly important with smaller henhouses because the overall volume of air inside will be relatively low. Because hot air rises, it makes sense for air holes or shuttered windows or vents to be positioned high up, just under the roof line. The presence of these is certainly a valuable feature to look for when buying. Some manufacturers get away with drilling a few small holes here and there, but in reality, this simply is not good enough.

Overcrowding should be avoided at all costs, especially if the birds are forced to spend long periods shut inside. It can promote all kinds of undesirable behavioral problems, such as feather pecking and egg-eating, and also increase the risk of health problems, including respiratory infections.

Most manufacturers of henhouses arrive at the capacity of their structures by calculating the number of birds that could roost side-by-side on the perch. This really is a theoretical maximum figure and should not be taken as a practical guide. It is like claiming that the back seat of a car will accommodate three adults. In theory, most will, but we all know it is much more comfortable with just two. Consequently, if you cut the recommended bird total by a third or even by a half, you will be doing your hens a big favor. It is not uncommon to see small houses, 3 x 3 feet (1 x 1 meter), being marketed as suitable for six birds. In reality, this limited space would be much better suited to three or four birds. So don't be taken in by overenthusiastic sales literature, instead make a reasoned judgement about capacity.

Space to move

Scratching about in a mixture of dirt and vegetation under a canopy of trees is just about as close to heaven as a domesticated chicken can get. Sadly, however, not many of us are able to offer our birds these free-range-type conditions, but we can try.

Out and about

Domestic chickens need to be let out of their enclosure every day, but what they are let out into will depend on the available space and your budget. At the bottom of the scale is the all-in-one house and run combination, bought by those keepers with a small garden or backyard. Some henhouses are designed with an integral run, but others have a separate structure of wire netting on a wooden framework that is butted up against the henhouse to create the same effect.

The add-on run units are best for the hens because they provide more space than the all-in-one alternative, but neither setup is ideal as a long-term arrangement, especially for larger breeds. In addition, unless you are

Above *Surrounding a run with electrified poultry netting can be the ideal anti-predator solution.*

Right *Chickens need as much space as you can give them to remain healthy and happy.*

able to move the henhouse and run at regular intervals, the birds will quickly transform the area inside into a muddy mess. Your options then will be to give up on the grass idea altogether and fill the run with a deep layer of wood chipping litter or give them a much larger area to play in.

A moveable feast

The traditional view is that keeping six laying hens in the best possible health requires a run area approximately 60 x 60 feet (18 x 18 meters), which is obviously out of the question for the majority of domestic keepers. What's more, that area used to be split into three equally sized portions, and the birds were rotated from section to section so that none of the ground ever became exhausted. Unfortunately, such an idealized setup is not workable for the average enthusiastic keeper.

However, if you do have the space, it makes a lot of sense to follow a similar principle. Divide the area you have into enclosed sections so that the birds can be moved from one to the other every few weeks. If you live in a wet area you might want to consider setting aside a section layered with gravel for winter use, or even building a large, covered run for when conditions get really wet and wintry.

Obviously, everyone's circumstances are different, and so most people inevitably end up with a compromise. The golden rule remains to give your birds the most space you possibly can.

Do it yourself

Many keepers prefer not to pay the often inflated prices being asked by commercial suppliers of henhouses and choose a more cost-effective option instead. Making your own enclosure, adapting a shed, or buying second-hand are all possibilities.

Keep it simple

The relative simplicity and sectional design of a standard henhouse means that building one should be within the grasp of anyone with basic do-it-yourself skills and tools. It will take time to do the job properly, however, and you will probably have to copy an existing design, drawing up your own set of plans in the process. The great advantage of this approach is that you can tailor your design exactly to your needs, taking the best bits from all the houses you have seen.

Other cost-effective options include buying second-hand or adapting a garden shed. The latter is cheap to buy and easy to convert for poultry use.

All you will need to do to get up and running is fit a pop hole, make up some nest boxes, fit wire over the window if there is one or drill ventilation holes if there is not, and install a perch or two. The height of most garden sheds will mean that the structure will contain a good volume of air. With the window open and a few additional holes drilled high up at the top of the walls, good ventilation should be assured.

Above *An older style pent-roof henhouse.*

Adding a wire door is great for damp days because it allows extra light for the birds but keeps them dry. You can simply close the solid door when it gets dark. On hot nights, you can leave the solid door open so that the wire door lets in air yet keeps the birds secure.

How big?

The size of the henhouse your birds will need is dependent on a number of factors—how many birds, their size, the space available, and your budget. As a guide, a simple pent-roof-style henhouse measuring 4 x 3 feet (1.2 x 1 meters) will be suitable for four or five large fowl or six or seven bantams to sleep. As a guide, allow a minimum of 3 x 3 feet (1 x 1 meter) for each bird. Thus, a henhouse measuring 6 x 4 feet (2 x 1.2 meters) could happily accommodate eight birds.

Where?

Ideally, the henhouse should have some shade from trees overhead to give it protection from the sun. The pop hole should be turned away from the prevailing wind, so that rain doesn't blow directly inside. Finally, it can be helpful if any side windows or vents are angled to face the afternoon sun, so that the warming benefits of this can be enjoyed during the winter.

Unfortunately, not everyone will be able to meet all these conditions, so most will end up having to make the best of their circumstances.

Security

If you choose to let your chickens range freely in your backyard, it is likely that you will need to think about fencing to keep them in and predators out. Never assume that just because you live in an urban area, you won't have problems.

Boundaries

Predators, especially foxes, represent a very real danger to chickens. Whether you live in a rural or urban setting, your birds will need protecting. Traditionally, this meant putting up a heavy-gauge wire mesh fence, about 6 feet (2 meters) high, with braced solid wood posts and wire tensioners. This would enclose the whole area, and the wire mesh would be dug in to a depth of at least 12 inches (30 cm) all around to guard against burrowing attacks.

Such precautions will not necessarily be enough to dissuade a determined predator like a fox, and many keepers supplement the fence with electric wires. These run all along the top and bottom of the fence. Both electric wires are positioned 8–12 inches (20–30 cm) away from the fence, with the bottom one set 8–9 inches (20–23 cm) above the ground.

This level of defense is a big investment and so tends to be something that keepers consider as they get more involved in and committed to the hobby. Of course, seeing the aftermath of a predator attack will alter your priorities dramatically; it is extremely upsetting for all concerned.

Shock tactics

Electrified netting is another option and is quick and easy to erect and simple to reposition when you want to change the location of the run. Typically, it is the horizontal strands that carry electricity, with the current being pulsed around the circuit by a single transformer, which can be powered from the main supply or a 12-volt battery.

Above *The fox remains the poultry keeper's number one natural enemy. Foxes are clever and will kill indiscriminately.*

The advantage of this is that foxes seem to hate it. Popular belief has it that they are able to sense the current and that provides enough of a deterrent. However, foxes can also detect when the current is not flowing, at which point they will be in like a shot. The control systems for this type of fencing are usually very reliable and, when problems arise, it tends to be a result of human error.

Often people will switch off the fence when they go in to feed their birds, then forget to switch it on again afterwards. These systems can also be prone to short circuiting, caused by branches touching from above or contact from vegetation growing up below. Because of this, users need to watch for stray branches and to keep grass trimmed back at all times. It is useful to keep a tester handy so that you can check the fence every day.

Rodents and other undesirables

One of the few unfortunate consequences of keeping poultry is that it increases the likelihood of attracting rodents to your property. Although we all know that rodents are everywhere already, nobody actually likes to see them.

Rodents

Rats and mice naturally find homes in compost heaps, log piles, and in old sheds. But the presence of poultry makes life particularly easy for them because it usually means a ready supply of food and a warm place to live. Spilled poultry feed and henhouses that sit low to the ground provide an irresistible opportunity to one of the world's great survivors. However, it is important that poultry keepers do all they can to discourage rats, because they can pose serious problems for both humans and chickens.

The fact that most poultry houses have wooden floors means that it is easy for rats to gnaw their way inside in a matter of minutes. Once inside, they will scavenge whatever food is available as the hens roost and may also chew off a few tail feathers just for the fun of it. The under-house environment is an ideal one for rats to colonize, and they will do it surprisingly quickly, given half a chance. Typically, a female rat will produce six litters a year, each of up to 12 babies, so colonies are rapidly established. With this sort of reproduction rate, things can very quicky get out of hand.

Above *Rodents are attracted by poultry and everything that goes with it. They cause nothing but trouble in henhouses.*

Above *Squirrels can also be extremely destructive.*

Right *Fencing the run securely is vital if you are to keep your birds in and all predators out.*

Catching Weil's disease (leptospirosis) is just about the most serious consequence we face from living in close proximity to rats. It is a serious disease that, if left untreated, can be fatal. You can catch it by allowing an open wound to be contaminated by rat's urine, which is why it is so important to wear gloves when you are clearing rubbish from an old shed where you suspect there may be rats. Rat urine will also contaminate water, and contact with this is another way of contracting the disease.

Swift action

If you suddenly notice gnawed holes in the floor of your henhouse or enlarged gaps around the bottom corners of a pop hole, suspect the worst and take action. You can either deal with the problem yourself, using poisons or traps, or you can call in a professional rodent specialist to do the dirty work for you. The important thing is to nip the problem in the bud and minimize the risk to all concerned.

There are plenty of different poisons on the market, any one of which will do an effective job if used correctly. But there are obvious risks attached, so you must work safely. Some of the products are more toxic than others, and they can pose a serious threat to pets, wild animals, and even children. Follow the product instructions to the letter, and do not take any chances.

Using traps is generally safer and, as with the poisons, there are plenty of styles to choose from. Some will catch the rodent live, others won't—the choice is yours. The problem with the live-catch type is that you're then left with the potentially gruesome task of disposing the animal yourself. Because this is not a pleasant job, plenty of people opt for the spring-loaded traps instead. You can even buy pre-baited versions nowadays that make life even simpler—for the operator, not the rat!

Keeping rat traps permanently baited will offer a good defense, but so will being careful with the storage of feed and avoiding needless spills.

Caring for Chickens

In much the same way that children are dependent on adults for ensuring that they eat a healthy and nutritious diet, chickens rely on their keepers to provide an adequate and correct supply of food.

Food and water

Feeding is a key aspect of good poultry husbandry, and whether your birds are to be kept on a free-range basis or live their lives in a small run, you are responsible for striking the right balance with the food they eat.

The digestive system

To start with some basics, chickens are omnivores, which means they will happily eat both vegetable matter and animal protein (in the form of insects, worms, slugs, and so on). However, because they have a beak instead of teeth, they cannot chew their food, only swallow it. Consequently, their digestive system is very different from ours. Food passes down the esophagus and into the pouch-like crop, which lies at the base of the neck (on the left-hand side). The food is held there while it is being softened by the addition of water, which causes it to swell. Feeling the crop is a good indicator of whether or not a bird is feeding.

Once it has sufficiently softened, the food continues on down the esophagus toward the stomach. This is actually a two-part organ. First, the food passes into the proventriculus where it is attacked by digestive juices that begin breaking down the protein content.

Left *Chickens are creatures of habit, so it is important not to change feed types or feeding methods.*

Above *Because chickens do not have teeth they cannot chew their food, which limits what they can eat. This startling Swiss Appenzeller has a horn-type comb and crest.*

The digestive system of a chicken

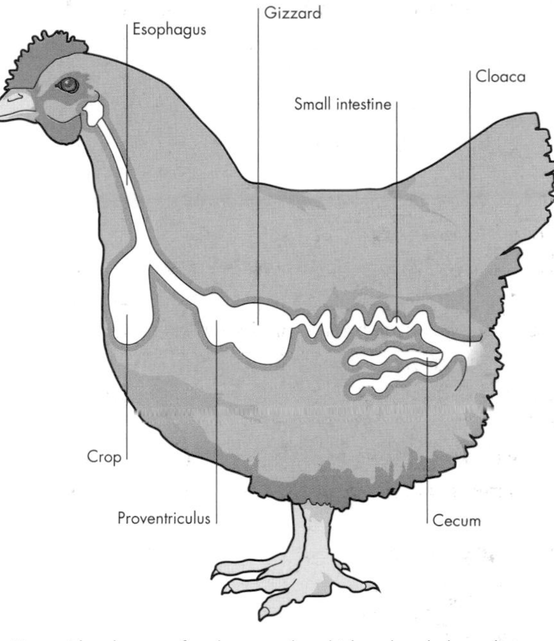

Above *The absence of teeth means that chickens break down their food by a combination of mechanical and chemical actions in their digestive systems.*

The second part of the stomach is called the gizzard, which is strong and muscular and uses mechanical action to break down the remaining food.

This process is helped by the presence of grit, which acts like tiny grinding stones to break up the hard constituents, such as whole grains and seeds. Amazingly, the food can be moved backwards and forwards between these two parts of the stomach until it is adequately broken down to be moved on.

Enzyme action

From the stomach, the finely ground (almost liquid) food is moved into the small intestine, where it is mixed with bile and digestive enzymes. The bile neutralizes the acids from the stomach and works on the larger fat molecules, while the enzymes work on the proteins, sugars, and smaller fat molecules. Most absorption of foodstuff takes place in the small intestine. Then, once it has passed through this section and into the large intestine, water is reabsorbed into the body.

The spent waste is then propelled through the remainder of the gut by waves of muscular contraction, toward the cloaca, which is the single opening at the back of the bird through which waste (droppings and uric acid) and eggs pass. A typical chicken will defecate between 25 and 50 times a day. What's more, you can tell a lot about the health of the bird from the appearance of its droppings. When all is well, it should show two distinct parts, a firm greenish-brown part, which is the feces, and a white, pasty part.

A balanced diet

The domestication of poultry has diminished a chicken's natural instinct to forage, and this, combined with restricted conditions and the laying demands we place on the birds, means that the way we feed them has never been more important.

Formulated feed

Fortunately, there is a neat, all-in-one solution available to us all: poultry formulated feeds. These commercially produced products are available in various forms to suit different ages and types of birds, but all contain a carefully controlled and formulated mixture of proteins, fats, calcium, vitamins, and trace elements—the essentials of life. The whole lot is mixed up and is available either as a mash or in the form of pellets. The important thing is to feed the right type to your birds at the right time. Options include formulations for chicks, growers, layers and breeders, and for birds you want to fatten for the table, and your choice will depend on what you want from your birds.

Of perhaps even more importance is something called chick crumbs, which is essentially a finely ground version of the same thing, with a particle size suitable for very young birds. Chicks hatch with an internal food reserve, and this will keep them going for two or three days. It is designed to tide them over while they learn to feed for themselves. To do this in the domestic environment, they must be presented with the correct food from day one, which is where chick crumbs come in. The young birds should normally be fed on this for the first five or six weeks, after which they can be gradually switched to growers pellets or mash, which contains slightly less protein and a bit more fiber. It will take the growing bird up to its point-of-lay (POL), the time the female starts to lay eggs, which usually happens at about 18 weeks old.

It is important not to overfeed the birds during this period because, if they become fat, it can lead to all sorts of nasty problems (including prolapse) once laying starts. As a guide, work on the basis that each bird should receive 4½ oz (125 g) of feed a day.

When birds are five months old, the feed type will need to be switched again. If you simply want eggs, offer layers mash or pellets, but if you want to breed with the birds or ultimately to eat them, you should opt for a breeder or a fattener formulation, as appropriate, although these formulations can be difficult for the hobby keeper to obtain.

Left *Types of poultry feed (from left to right): chick crumbs, pellets, and mash (meal).*

The traditional approach

Of course, pelleted food hasn't always been available, and many experienced keepers still don't use it. They prefer to stick with the traditional alternative, known as mash (or meal). This offers the same basic mix of ingredients, but in a loose form. It can be ground to different degrees of coarseness and fed either dry or wet (mixed with a little water, to a crumbly—but not sloppy—consistency). It is slightly cheaper to buy than pellets, and many breeders claim that mash gets digested better by their birds. On the downside, though, it can allow the birds to feed selectively—to pick out all the tastiest bits and leave the rest.

No matter what type of commercial food you give your birds, always check the sell-by date carefully. The vitamin and mineral content in feeds does deteriorate with time, and the general thinking is that it is best not to use anything that is more than three months old. In addition, pellets and mash must always be stored in a sealed container to protect it from rats and mice and to keep it completely dry so that it does not go mouldy and turn sour.

True grit

One thing that no poultry keeper must ever forget is the vital role played by grit in a chicken's diet. Its presence in the gizzard is essential to effective digestion. Left to his own devices, a bird will pick up plenty quite naturally as he pecks in the earth, but birds that are not lucky enough to enjoy a free-range life will need a grit supplement. It can be bought in all sorts of ways, but a good-quality mixed grit is one of the best. This is likely to contain granite fragments plus crushed oyster shell, which is a good source of extra calcium and is needed by the hen for the production of eggshells.

Make sure that a supply of grit is always readily available to your birds; a small container of it inside the henhouse is ideal.

Left *Chickens need a regular supply of grit to aid their digestion. Birds allowed outside will find their own, but it must be provided for all birds that are not kept on a free-range basis.*

Leftovers and treats

Chickens can certainly be fed, and will enjoy, most natural scraps from the kitchen—leftover fruit, vegetables, corn cobs, peelings, and so on. But remember, if you are offering good-quality pellets, nothing much else should be necessary.

Above *Birds enjoying a balanced, pellet-based diet shouldn't need much else, but they do love an occasional treat.*

Kitchen scraps

Potato skins should be boiled in unsalted water first—other green vegetables can be prepared in this way as well if you wish—and then dried off with mash. Fresh green vegetables, such as cabbage, lettuce, and spring greens are also always popular—cooked or raw.

The foodstuffs to avoid are those that are salty or sugary, because they can be harmful if too much is eaten. Some keepers like to feed their birds treats, like bread and butter, yeast spread on toast, and cereal. You won't find many official sources recommending any of these (although plain cereal is fine as a healthy treat), but owners do it anyway because their birds appear to like it. The best advice is to avoid feeding them if you can, but, if not, to do so only very sparingly.

Chickens love mixed corn, maize, and wheat, but take care that you do not offer too much because they can be fattening. These grains are great as a daily treat, however, and if you scatter a handful or two in their run for them to scratch around and find, it will keep them amused and contented for hours.

Additives and supplements

Lots of companies sell all sorts of wonderful-sounding additives and supplements that are supposed to bring this and that benefit. Unfortunately, there is little documented proof that any of these "elixirs" actually do much good. It is probably best not to bother with most of them.

However, there are a few more natural options that are worth considering, although it would be sensible to discuss them with your vet before proceeding. One of the oldest and best-known additives is apple cider vinegar (ACV), which is claimed to offer all sorts of

Left *Aloe vera has natural antibiotic, anti-inflammatory, and antiseptic properties and is said to boost the immune system. Many poultry keepers claim that it is beneficial for their birds.*

health benefits to poultry. The use of vinegar as a medicinal cure-all goes back centuries, so at least the product has some medical credibility behind it. Apples, too, are full of natural vitamins and minerals, and the two together provide a tempting prospect.

Suppliers will tell you that ACV works by slightly increasing the acidity of a bird's body, thereby making him less likely to suffer from problems linked to bacteria and viruses. Some people also claim that it is a valuable tonic, helping birds through stressful times, such as the annual molt, when natural energy levels are low. If you decide to try it, do make a point of buying unfiltered ACV, not the clear product you see in bottles on the supermarket shelves. This is a cooking ingredient and nothing more.

Garlic is another potentially useful natural addition to the diet, and many experienced breeders also say that they use aloe vera, which has natural antibiotic, anti-inflammatory, antiseptic, and immune system-boosting properties for all sorts of useful purposes.

Above *A wide range of feeders is available. Make sure that what you use is suited to the birds you have.*

Feeders and drinkers

Give a little thought to the containers that you use to hold feed and water for your birds. You can spend a lot of money or very little, but what you do will be determined by your setup.

The simplest feeder is a straightforward bowl. An earthenware one is best because light plastic bowls can be tipped over. Your choice will depend on the number of birds you have and whether you are feeding pellets or mash. Both can be given in a bowl, as long as bird numbers are low. However, there is a risk that the food will become contaminated as the hens walk all over it. Custom-made feeders are more expensive but are probably more satisfactory in the long run.

It is best to position feeders under cover—hanging them from the roof in the henhouse or covered run area is ideal. Some types have large umbrella-like covers, and these are intended for use in the open. You can buy all sizes in plastic or metal, and the choice really is up to you. The worst thing you can do, though, is buy one that is too small, because you will be forever filling it up. Troughs with inward-curling edges or built-in grills are a good choice because they help reduce the amount of food that gets flicked out by the birds.

The same general rules apply to drinkers. Bowls can be used for adult birds—but not for chicks, which might drown. Large-capacity, purpose-made drinkers are better, especially if the birds have to be left for a few days. Running out of fresh drinking water is one of the worst things for chickens, and many novice keepers underestimate how much birds drink. As a guide, a large drinker that contains 6½ gallons (30 liters) will last eight birds for about two days.

Bear in mind also that if you chose to keep crested breeds, such as the Poland, you will need to buy special narrow-lipped drinkers to prevent them from getting their crest feathers wet as they drink. When you buy these rarer breeds, ask the advice of the supplier about suitable food and water containers.

Get a grip

Learning how to handle a bird is one of the essentials of good husbandry. It is important that both you and the birds get used to it, because regular hands-on inspections are useful for detecting weight loss and the presence of lice, mites, and other parasites.

Proper handling

It is important to handle your birds correctly, because getting it wrong will cause birds to panic and become stressed. And stressed birds are bad news—they are much more likely to stop laying and may even pick up infections. There is a definite method to be followed.

The first thing you need to do is make sure that your birds are comfortable in your presence. If you have reared them from chicks and been handling them from day one, it is hoped that they already are. But if they are new to you and you bought them as adults, they probably aren't. So establish a regular feeding routine and remain with them while they eat, standing still at first. They will soon get used to you.

Catching a bird

You must adopt a calm and gentle approach to actually catching a bird. Sudden movements, loud noises, and lots of other people watching are all certain to lead to disaster. It is a job best done at dusk, when the birds are calm and preparing to roost. Gently usher the bird you are after into a corner, adopt a crouching position, and spread your arms wide to discourage him from making a run past you. Once you and the bird are settled, make a confident and decisive grab for his legs. Never reach for the neck, wings, or tail.

Once you have hold of the legs, draw the bird into your chest to quell the inevitable flapping and calm him. All chickens will sit happily if they feel balanced, and the best way to do this is to support them from underneath. Extend your forearm and hand down the length of the bird's body, from front to back. Have your hand palm-up, with your fingers spread so that his legs interlock between them. Loosely grip the legs, making sure that you have one finger between them so they are not crushed together.

Held in this way, balanced with his breast along your forearm, the bird will instantly feel safe and relaxed and will be happy to sit there for as long as you want. On a practical level, another good reason for holding a bird this way around (with his head facing behind you), is that if he should mess, it will drop harmlessly in front of you, rather than all over your arm and legs.

Above *Regular handling will allow you to inspect your birds for signs of trouble, including nasal discharge and parasite infestation.*

Just about the worst things that you can do are smother the bird or hug him tightly. Both are guaranteed to instill panic as far as he is concerned, at which point you will have lost the contest and probably traumatized the bird. Some breeds are naturally more difficult to handle than others. Those with a reputation for being "flighty"—the lighter breeds such as the Leghorn and Ancona—will always tend to be a handful, even for more experienced keepers. Heavy breeds, like the Orpingtons, Sussex, or Rhode Island Reds, which are known for their docile characters, will be a good deal more keeper-friendly, assuming, of course, that you do things properly.

If you do not manage to get a hold of the bird after the first couple of attempts, do not go on trying. Repeated failures will simply upset you and the bird and destroy any trust that had built up between the two of you. Walk away, let things settle down, and try again another day. It's all a matter of confidence.

Above *Balanced support is the key, as with this Hamburgh bantam.*

Keeping it clean

Cleanliness is crucial. All poultry keepers should understand that a henhouse that smells of ammonia or is visibly strewn with encrusted droppings on the perches and floors poses a serious health risk to the birds.

Henhouse hygiene

Keeping chickens in overcrowded conditions must be avoided. It is bad news for the birds and, ultimately, bad news for you. Apart from all the problems posed by heightened stress levels, birds that are packed too tightly into a henhouse are also likely to become bad tempered and aggressive with one another. This, in turn, can promote destructive behavior, such as feather pecking and egg-eating. Both of these can be easy to start, but very difficult to stop. Chickens will quickly fall into bad habits if they are allowed to do so, and it is up to you to make sure that this does not happen. The other consequence of an overcrowded henhouse is that it will need cleaning far more often than one that is correctly populated, so establishing the correct number of birds is essential.

Above *Dust-free litter materials are best for your birds. Use them in nest boxes and on the floor.*

Some henhouse designs include a droppings tray that slides in on the floor below the perches. It is put there to catch the droppings produced while the birds are roosting at night (about 50 percent of the output) and must be cleaned off every day or so, otherwise it will be a haven for mites. Droppings should never be allowed to build up to any degree anywhere in the enclosure. Similarly, the litter spread on the floor and in the nest boxes must never be left so long that it becomes damp, smelly, or moldy. Any one of these conditions will greatly increase the risk to the birds of disease, particularly respiratory problems.

Floor coverings

There are several materials that can be spread on the floor of your henhouse. Traditionally, straw, which was both cheap and readily available, was the favored option. It is, however, quite a high-maintenance choice. The danger is that liquids can soak through and start to fester within the layer, while leaving the top surface looking perfect. To avoid problems arising, straw needs to be turned regularly so that you can keep an eye on the degree of contamination.

These days many experienced keepers choose to use wood shavings as a cost-effective and practical litter. This can be used in the nest boxes as well as on the floor, but it is important to specify light-colored, softwood shavings only. Sometimes, soft and hardwood (dark-colored) shavings are mixed, and this is not suitable because hardwood produces splinters.

Sawdust is another wood-based alternative, which some keepers are tempted by because it is cheap.

Right *Henhouse floor litter must be kept clean and fresh at all times.*

However, the big risk with this type of litter is that it brings dust into the henhouse. This is a real danger for your poultry because it actively promotes respiratory problems. To counter this risk, a number of newer bedding materials, some of which are based on hemp, have been adapted from the horse market. These represent the most expensive option but, being completely dust-free, are ideal for chickens and minimize respiratory risk.

Sweeping clean

It is important to be methodical about keeping your henhouse clean. Establish a routine and stick to it if you want to guarantee the best conditions for your birds.

Ideally, perches should be removed from the henhouse, scraped clean, and then hand-scrubbed with a livestock disinfectant every two weeks or so. Obvious signs of localized contamination should be removed from the floor litter whenever they are spotted. Overall, though, the whole lot should last up to 12 weeks before it needs to be completely changed, although this will depend on the number of birds you have.

You will have to judge for yourself, based on a balance between how the litter looks and smells and its affordability. Remember that changing the litter too frequently will do nothing but good, but leaving it in place for too long is almost certain to cause trouble.

You need to keep an eye on the litter in the nest boxes, too. It is a good idea to clean and disinfect the boxes every couple of months (more frequently in hot summer weather) and to change the litter as necessary.

Once a year you will need to give the whole henhouse a thorough cleaning, preferably using a pressure washer. Do this when the weather is warm to minimize the drying time and, after washing, use an agricultural-type disinfectant (not a domestic one, which won't be strong enough) to scrub all surfaces thoroughly.

Keeping a proper run

If lack of space or other circumstances means that your birds have to spend most of their time in a run adjacent to their house, the condition of that area is of paramount importance and it must be kept clean.

Looking after the run

What may start as a lush, grassy enclosure can quickly be reduced to a dirty, muddy mess, even by a small group of birds—this is known as fouling the ground. Dealing with the problem properly involves moving the birds to another area, while the original patch is given time to recover, after being leveled and reseeded. But this obviously requires even more space and so is not possible for many keepers.

A practical alternative can be to forget about the grass altogether and instead to create a graveled or deep-litter run. The birds will be just as happy—providing you get the conditions right—and you will have none of the hassle associated with trying to

Left *Giving your birds space and plenty to interest them inside the run will help keep them active and healthy.*

Above *It is vital to keep poultry active to avoid the risk of the birds becoming overweight.*

resurrect flagging vegetation. This approach is particularly suited to smaller runs and those that are covered with solid roofs and have walled sides.

The level of exercise and amusement of chickens kept in relative confinement need special attention. Exercise is important because well-fed domestic poultry can easily become overweight if they are not able to scratch around in a free-range environment, as nature intended, and this affects their general health and their laying performance. It is important that they are encouraged to work for their living.

Mixing a small amount of pellets or corn into the gravel or litter will make them scratch about with added vigor to find it. You can also experiment by placing different obstacles in the run and varying the layout every now and then. Suspending a fresh cabbage or lettuce in the run at just above head height will make the chickens jump a little as they peck, which is good exercise for them, especially in a small run.

If the run is covered and well protected from rain, you can use softwood shavings as your deep-litter material. However, if there is a chance that rain will blow in or if there is no roof, you would do better to opt for gravel or bark chips. Whichever material you choose, it will usually stay fresh for about a year, after which it will need to be dug out and replaced or replenished.

It is all too easy to ignore or overlook the condition of the henhouse and run floor coverings. Always remember, though, that your birds are completely dependent on you for their health and well-being.

DUST BATHS

Birds that spend most of their time in an enclosed run should have access to a dust bath. This is nothing more than a shallow box that has been half-filled with dry, powdery earth or fine wood ash (do not use coal ash, which can encourage scaly leg). It should be big enough for the largest of them to spread their wings as they "bathe." Chickens love to take a dust bath and do so as one of their natural defenses against lice and other parasites. So it is important that they are provided with this little extra luxury

Above *Chickens need access to a dust bath to help keep parasites in check. It is nature's way of dealing with lice and mites.*

Preventive action

Watching how a group of birds interact and becoming familiar with their routines will teach you a great deal. In time, you will learn to spot the tell-tale signs of trouble, and this ability will enable you to deal with any problems quickly.

Worming

Taking time to observe your birds is a vital aspect of good poultry keeping, but there is more to it than just watching the birds. Handling is important, too, because picking up your birds will enable you to check for parasites and to assess their weight and general condition. Parasite-related problems and other poultry diseases are discussed later in the book (see pages 190–199), but here we need to concentrate on worms.

A number of worms can cause problems for poultry, but one of the most common is roundworm. This is among the largest to affect chickens—it is thick and can measure up to 4¾ inches (12 cm) long.

Above *Time spent watching your birds and how they interact is never wasted as it will enable you to spot anything unusual.*

Roundworm eggs are frequently carried by earthworms. Once eaten, the eggs develop into mature adults in just six weeks in young hens (eight weeks in older birds). These start producing eggs themselves, which are then carried out of the body in the bird's droppings and eaten by more worms, and so the process continues. Roundworm infestation can be serious for chickens, leading to weight loss, feather-fluffing, diarrhea, huddling, and, in the worst cases, death. Moderately infected birds will be slow to develop and gain weight and will appear generally slow.

Poor environmental conditions are a major factor in the likelihood of roundworm infestations, with boggy ground or even wet patches around badly sited drinkers being common causes because earthworms are attracted to damp conditions. As birds get older (beyond three months), their resistance to roundworm increases as changes occur in their intestines.

Tapeworms also commonly affect poultry, although the consequences are not quite as serious as with roundworm. The adult tapeworms bury their heads in the wall of the chicken's intestine, and parts of its segmented body break off once they are full of eggs. These are passed to the outside in the droppings, are eaten by snails and beetles, and are then carried into a new host when the insects are themselves eaten. Symptoms include weight loss, general fall-off in feather condition, weakness, and slow growth. In bad cases, keepers may notice bloody diarrhea.

Worms can be effectively treated using a specially formulated poultry wormer; ask your vet to recommend a suitable brand. This should be done at least twice a

Left *Claws will need regular clipping, particularly on birds that don't free-range.*

year. Bear in mind that the use of some worming products means that you will have to stop eating the eggs for a certain length of time. The duration of this withdrawal period will vary according to the product, and you should always consult your vet or the product manufacturer if you are in any doubt at all.

Beak and claw trimming

Beaks are formed from keratin and grow continuously on chickens to allow for wear and tear. Birds kept in domestic conditions and fed predominantly soft food tend not to experience the degree of activity needed to keep this growth comfortably in check. Therefore, as a responsible keeper, you must keep a watchful eye on this growth and trim back when necessary.

The upper beak (mandible) grows faster and, if left unchecked, will extend over and beyond the tip of the bottom part. If he is to peck effectively, the bird needs the two halves of its beak to be more or less matched in length, and if the top part is allowed to become longer, eating and drinking problems will result.

The section that needs to be removed is usually an obvious, lighter color. This can be carefully snipped off with straight-bladed nail clippers. Take care that you do not cut off too much and drift into live tissue. If you are worried about tackling this yourself, your vet or an experienced poultry keeper will be happy to oblige.

Claws will need regular clipping, too, especially if your birds are kept on soft litter and without much access to free-range conditions. As with beaks, it is not good to let claws grow too long, otherwise it becomes uncomfortable for the bird and can even lead to foot problems. This, too, is a job you can tackle yourself with care, or you may prefer an experienced pair of hands to do the job for you.

Cockerels develop an additional "claw" on the back of the leg, a short distance up from the foot. This is called the spur, and it will need trimming also.

BEAK PROBLEMS

Chickens can hatch with a variety of beak problems, most of which are caused by genetic defects. These birds should be avoided, particularly if you intend on breeding chickens. Common among these is twisted beak, when the beak grows noticeably to the left or right, rather than straight ahead. There is also a condition known as open beak, when a gap develops between the top and bottom beaks, even though the two come together at the front end. And there are instances when the lower beak is much shorter than the top one. All of these problems will give the bird difficulties with eating and drinking.

Above *Beak problems, like this poor alignment, will affect a bird's ability to feed.*

Fine feathers

A chicken's appearance is largely determined by his plumage. The shape and color of the feathers help set breeds apart from each other, but what they have in common is that all their feathers are changed once a year, in a process called molting.

Molting care

The molt, which usually begins in late summer, is a stressful time for the bird, even though it is a completely natural process. Feathers are actually about 80 percent protein, so replacing them is a major operation for a bird, and virtually all bodily resources are thrown into the task, which is why egg-laying stops for the duration and the bird can appear at a bit of a low ebb. The time taken to complete the molt varies from breed to breed, but a healthy bird in the first or second molt should take no more than eight weeks. Variation in the molting period among birds of the same breed is a good indicator of a laying performance: those that complete the process most quickly will be the best layers.

Chickens do not molt in their first year of life. For example, birds that hatch in spring will have developed a full set of feathers by the autumn, and their first molt will occur 12 months later. Molting gets slower as the

Left *Don't be alarmed by the way birds look during the annual molt. It is a completely natural, rejuvenating process.*

Above *Wing clipping is needed on the flightier breeds if you want free-range birds to stay put. Trim the tips off the flight feathers of one wing.*

bird gets older, and the period without eggs increases. This absence of eggs needs to be taken into account if you are hoping to maintain a supply of eggs during the winter months.

It is important that birds are well cared for during the molting period and that upheavals and changes to the normal routines are minimized. It is usually beneficial to supplement the diet to help meet the birds' increased requirements for protein. You can buy special "maintenance" feeds for this purpose, and some keepers like to feed a small daily ration of extra protein.

The neck molt affects young hens (pullets) that are just reaching point-of-lay. As the name suggests, this is a partial molt, limited to the neck feathers, but it is sufficiently draining on the bird to delay egg production. It is not a good thing and tends to occur on birds that are reaching POL too early because they have been switched from growers to layers pellets too soon. Neck molts also occur when a pullet starts to lay in early autumn and then stops as day lengths shorten.

Clipped wings

Clipping a bird's wings may sound cruel, but it is not. It is a technique used by keepers of the flightier breeds to ensure that they remain where they should be. Although no breeds are known as accomplished fliers, several of the lighter types can propel themselves far enough to get up into trees or over fences and hedges. This can be inconvenient for keepers—particularly those in an urban environment—who don't have the luxury of a completely enclosed run area.

To prevent escapes, it is the simplest of jobs to snip off the tips of the main flight feathers. This needs to be done to one wing only, and it will unbalance the bird if it tries to fly, causing it to flutter in circles rather than a straight line. However, as with beaks and claws, it is essential that you do not take off too much when clipping wings, otherwise the cut ends will bleed. Use the quill color as your guide—only ever cut in the light part, staying about 2 inches (5 cm) away from where the feathers leave the wing.

Above *It is very important not to neglect poultry during the annual molt, as the birds' stress levels run higher than usual at this time.*

Bad habits

Like humans, chickens can eat too much of the wrong food. Also, vigilant keepers should keep a wary eye out for feather pecking and egg-eating—bad habits that poultry can get a taste for and that must be stopped before they gain a hold.

Chubby chickens

Hens kept in the backyard can suffer from obesity in much the same way that humans do. Too much of the wrong food and not enough exercise are the common causes, and, unfortunately, it is we who are to blame. Overfeeding chickens is bad for them. They will eat what they are given, but if you give them too much, the weight can pile on and egg production will suffer.

Above *Docile breeds, like the Orpington, can suffer from overfeeding.*

In addition, obese males are likely to experience fertility problems. You should be particularly careful with the less active, docile breeds, which are more likely to succumb to the effects of over-enthusiastic feeding.

When you are handling your birds, make a point of feeling down to the breast bone. It should be detectable under a reasonable layer of fat. If it feels sharp-edged the bird is underweight, but if it is hard to feel at all the bird has gone to the other extreme. A second test for obesity is whether or not you can pinch (gently) any fat at the rear end. Both of these, of course, are subjective tests, so it is hard for beginners to be sure about what they are feeling.

What everyone can do, however, is control the amount of feed they provide. Ideally, there should be nothing left when the birds turn in to roost. Avoid refilling feeders during the day and try to stick to a set allocation of about 4½ oz (125 g) a bird each day. Things will find their own level after a week or two's careful observation.

Feather pecking

Birds usually peck at each others' feathers because they are bored or overcrowded. It has also been suggested that they are prompted to do it by the presence of parasites, such as lice, on the skin under the feathers, or because some aspect of their diet is lacking. Whatever the cause, it is behavior that you need to be alert for. Chickens can be spiteful, cruel creatures once they get a taste for blood, and it is not unusual for a lowly bird that is being picked on like this to be pecked to death.

Above *Feather loss like this, which exposes raw flesh, will attract other birds to do more damage.*

Stopping this behavior can be difficult. As a rule, it is best to remove and isolate the injured bird so the wounds can heal. If you can give the birds more space, do so; if not, provide extra amusement in the henhouse and run. Add some obstacles, hang some fresh vegetables, or even try suspending some old CDs on a string for them to peck at. All these additional diversions can help remove the temptation.

Egg-eating

Egg eating is another vice that can be difficult to cure. Hens will quickly get a taste for the contents of a freshly laid egg, but the question you must ask yourself is how the egg got broken in the first place. There are a number of possibilities, most of which relate to poor husbandry practice. Eggs that are left in the nest boxes stand more chance of being accidentally broken by another bird; it will then be eaten by successive birds that come into that box. So, the first rule is to always collect eggs as soon and as often as you can.

Overcrowding in the nest box can be another cause of broken eggs. If the box itself is too large—so that more than one bird at a time can fit inside—this will encourage the problem. Also, if one of the nest boxes is more attractive to the birds than the others, it will be used more and shells may get cracked. Consequently, it is best to have all nest boxes be identical and fixed at the same height inside the henhouse.

Other theories suggest that lighting can play an important part in broken eggs. It is said that if it is too bright in the nest boxes, egg-eating is more likely to occur. A piece of hessian pinned across the opening of each box can help in this respect.

Finally, "soft-shelled" eggs (sometimes produced by young or fat hens) can provide a temptation that is just too great to resist. These eggs, which are laid without a shell, are produced every now and then by most hens. But frequent occurrences point toward a more serious problem, possibly a calcium deficiency in the diet (make more oyster shell available) or something wrong internally with the bird. If the problem persists, seek veterinary advice.

Above *The damage to the comb on this Minorca hen is a sure sign that the bird is being bullied. It may also be suffering from feather pecking.*

Hatching and Rearing

Once you have spent the first few months with your new birds and have established a husbandry routine with which both you and they are happy, your thoughts are almost bound to turn to increasing the size of your flock. Poultry keeping affects you that way: it is a totally addictive hobby, and there are so many new avenues to explore that expansion is almost impossible to resist.

More birds

The simplest way of obtaining new birds is to buy stock from a reputable breeder. But for the more adventurous among you—and those with young children—hatching and rearing your own chicks can be an exciting and educational way to progress.

Hatching options

Essentially, there are two choices: the natural and the not so natural. Both have their pros and cons, and your final decision will be based on the space you have available, your budget, and the type of birds you already keep.

Chickens are amazing creatures in that they possess the urge to go on laying eggs, regardless of the season. Most other birds and animals tend to limit their procreational activities to a set time of year, usually spring. But the chicken carries on producing eggs virtually the whole year round, although there is a break during the molt and production rates do slow down during the dark days of winter.

Much of this performance, of course, is the result of the careful selection process that was carried out by knowledgeable breeders in the past. The hybrid breeds represent the pinnacle of this "art," and the best of these birds are capable of producing more than 300 eggs in a year, which is an astonishing feat.

It is worth noting that every hen hatches with a finite number of egg cells already in her ovary, and this governs the total number of eggs that that bird can produce—most, in fact, never manage to reach their theoretical maximum. The greatest number of eggs will be produced in the hen's first laying season, but she will lay larger eggs in the second year, albeit slightly fewer of them. There will be a significant fall in numbers during years three and four (commercial laying hens aren't kept for longer than two years), and laying performance will continue to fall throughout the rest of the bird's life. A healthy, well-managed hen will happily live to the age of nine or ten, but egg production might well have dried up altogether by the time she reaches this stage.

Nature's way

In the wild a hen will continue to lay an egg a day until she has produced a manageable clutch, which can be anywhere between 6 and 15 eggs, depending on the breed. At this point, laying will cease and the hen will go "broody," meaning that she will sit on the eggs, providing the heat needed to develop the embryos within. This takes 21 days for large fowl (a day or two less for bantams), and during this period, the broody hen will regularly turn and reposition the eggs under her so that all of them get just the right amount of heat from her body.

When kept in a domestic environment, it is this desire to produce a clutch of eggs that keeps the hen laying regularly—as long, of course, as those she does produce are removed. Unlike many other species, a hen will happily go on producing a regular supply of eggs without the stimulus created by the presence of a male. If you want to raise chicks, however, the eggs obviously need to be fertilized, and for that your hens will have to run with a cockerel.

Alternatively, you can simply buy some fertilized eggs from a recommended supplier.

Going broody

It is impossible to predict if a hen will go broody, but if you have chosen an appropriate breed, the chances are that she will.

The maternal instincts needed for a hen to turn broody are much stronger in some breeds than in others. Hybrids do not make reliable sitters, nor do many of the light, flighty breeds, such as the Leghorn or others from the Mediterranean region. The best broodies will usually be found among the heavy, docile breeds, including the Cochin, Brahma, Sussex, Orpington, Rhode Island Red, Plymouth Rock, and Wyandotte. Others worth considering include the Yokohama, Sumatra, Scots Grey, and Old English Game. Silkies are among the best-known breeds for going broody, which they do reliably and often. The reliability aspect is important because the last thing you want is a hen that changes her mind halfway through.

If the hens have not laid the eggs you want them to hatch, you will need to encourage them to go broody. This can easily be done with good sitters by placing six fake eggs, golf balls, or round stones in the nest. It is best to put the hen and the fake eggs in a separate sitting box and to place them in a quiet, predator-proof area away from the other birds. An old shed is ideal. It should also be out of direct sunlight if the weather is hot. Many keepers also add a little maize to the hen's feed ration at this stage.

The whole process might take two or three weeks, so be patient. This is why you should not use real eggs right from the start—they are quite likely to get broken over that sort of period, which will simply give the bird a taste for egg-eating. They will probably go stale, too.

You must be guided by a number of signs when you are deciding if a hen is actually broody. When she turns broody she will begin spending a lot more time

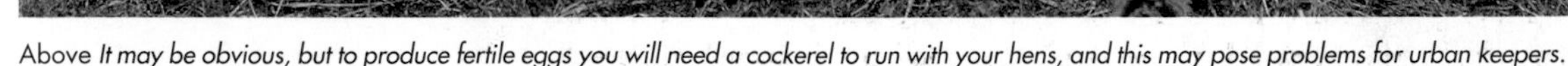

Above *It may be obvious, but to produce fertile eggs you will need a cockerel to run with your hens, and this may pose problems for urban keepers.*

in the box and will start raising her feathers when you get close. If you lift her out of the box, she will make a characteristic "clock, clock, clock" sound, and you should notice a loss of feathers from her breast area. She may also spread her wings when placed on the ground (appearing quite aggressive), and if she stays in her sitting box at night you know for sure that she has gone broody.

If you are hatching your own birds' fertile eggs, it is important that they are fresh when they go under: no more than seven days and not less than 24 hours old.

Sitting comfortably?

You must not forget about the hen while she is broody. She will need to be lifted out of the box for 20–30 minutes every day so that she can feed, drink, and defecate. This can be done in the morning or evening—whichever suits you better—but it is important to get the routine established up to a week before you place the real eggs. Clean the eggs with a special egg disinfectant before placing them under the bird. A large hen should be happy to sit on 10–12 eggs, while a bantam can be expected to manage six or eight.

Above *Once broody, the hen should be placed on a nest in a separate sitting box like this: a commercial unit with two compartments.*

Many beginners are disappointed when birds desert the nest prematurely. For natural brooding to be successful, it is vital that the hen sticks to her task, and to do this she must be both genuinely broody and comfortable. The comfort aspect is one that is often overlooked, but this can be a costly mistake—get it wrong and she will not settle.

You must make sure that the hen is free from fleas and lice, because this sort of parasitic irritation could drive her to distraction during the long sitting hours, causing her to give up. Give her a thorough dusting with a good-quality anti-lice powder right at the start of the whole process.

The environment is important, too. The sitting box should be peacefully located so that the bird feels secure and relaxed. Many commercial sitting boxes do not have a floor, and the straw nest inside sits directly on the ground. The idea is that moisture from the earth will keep the eggs in a humid atmosphere, which is important for their successful development. The hen should be fed straight wheat and fresh water during the sitting period, and she should have access to grit. She will also appreciate a dust bath.

It is important that the hen defecates while she is on her daily break. It is not good for her to soil the nest because this can contaminate the eggs; but if it does happen, you should carefully remove the dirty nesting material and replace it with fresh.

If you don't have a floorless sitting box that is sited on the earth, you will need to recreate the humid conditions necessary by sprinkling each egg with just a few drops—no more—of water at body-temperature at the end of the hatching period (on days 18, 19, and 20 for large fowl and 16, 17, and 18 for bantams). Always do this just before the hen returns to the nest after her break, otherwise you risk chilling the eggs.

Hatching day

When hatching day arrives, keep disturbance to a minimum. If you are lifting the hen to check on

Above *Experienced breeders run two broody hens at a time so that all is not lost if one deserts the cause.*

progress, do so gently and be careful that there are no chicks tucked under her wings as you lift.

Even after hatching has started, you will need to lift the hen off the nest for a break. When you do, gently lay a small blanket over the nest. Not only will this keep the chicks warm, but it will minimize their cheeping, thus allowing the hen to have a decent break without being called back to the nest before time.

Take care that partly hatched eggs are not broken when the hen is removed or resettles on the nest. Some keepers prefer to remove them while the hen is being moved to avoid the risk altogether. Replace them when the bird is back in place, with your hand shielding the egg in case she is tempted to peck at it. As the chicks hatch, their discarded shells should be removed to avoid them accidentally encasing another, unhatched egg. Any eggs that don't hatch should be disposed of.

Once the chicks have hatched and "fluffed up," they and the hen should ideally be transferred to a broody coop containing a litter of wood shavings; straw can be difficult for very young birds. Include a shallow dish of chick crumbs; the hen will teach the youngsters to eat. She will also scratch some of the food into the litter, which is quite normal. They will need water, too, but make sure you use a proper chick drinker so that there is no danger that they will drown.

Keep the hen confined in the coop until the chicks are at least a week old, but if the weather is reasonable, allow the chicks into a small run beside the coop. The hen can be allowed in there after a week. When the chicks are three weeks old, they will need a larger run. Hen and chicks must be shut in every night. The hen should stay with the chicks for six to eight weeks, after which she can be returned to her normal pen with the other adult birds. She may well start to molt at this point, which is quite normal.

Do not let the hen and chicks have a large area to run for the first week because the young chicks tire easily, and there is a danger that the chicks will be left and get chilled.

As with many aspects of poultry keeping, being methodical and confident are the secrets to successful natural brooding. It is important to establish a routine with your broody hens, and to stick with it.

Above *The hen and her chicks should be transferred to a broody coop once the youngsters have fluffed up. The chicks can explore the run if the weather is good.*

Using an incubator

Hatching eggs with a broody hen may be the way Mother Nature intended, but it is not the most practical option for a lot of keepers who have neither the space nor the facilities. If you fall into this group, an incubator offers a more convenient alternative.

Types of incubator

These machines, which come in all shapes and sizes, provide a carefully controlled environment in which eggs can incubate and hatch, and they are designed to replicate the sitting hen in terms of heat and humidity. The smallest, budget-priced units, which simply use an electric lightbulb as the heat source, are known as still-air machines. The more expensive units have heating elements and internal fans to ensure that the correct temperature is retained throughout the unit. These are referred to as forced-air incubators.

The other essential requirement is that the eggs are regularly turned during the incubation period—this is vital for successful hatching. Each egg (lying on its side) must be rotated 180 degrees at least three times (preferably five) every day. Always pick an odd number of rotations to avoid the eggs ending up in the same position every night. Eggs must be turned in this way for the first 18 days of the incubation process. Forgetting to do this is one of the biggest causes of hatching problems for beginners who are using a basic incubator. Marking the shells with a cross on one side will help ensure that you turn each egg by the right amount each day.

The more expensive machines will have built-in turning mechanisms (semi-automatic or automatic). With the entry-level machines, the operator is left to do the turning by hand, which is why these machines are known as manual incubators.

Try, try again

Success with an incubator cannot always be guaranteed. Certainly, if you are using one of the cheaper, less sophisticated machines, you might have to experiment

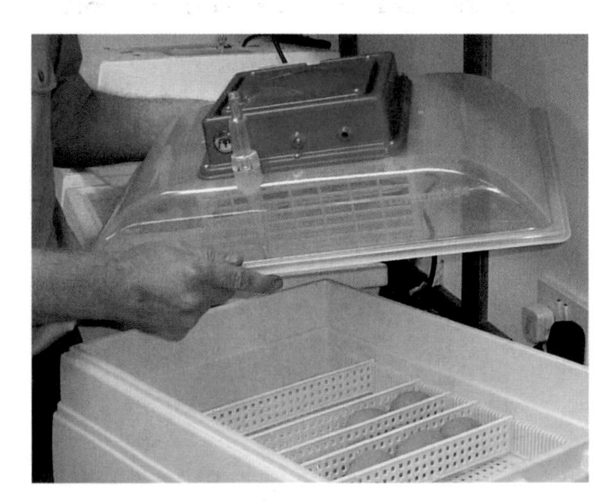

Far left *Larger incubators have more features and greater levels of control, but you will pay for the convenience with a higher purchase price.*

Left *Inexperienced keepers should start on a small scale, with a four- or six-egg manual incubator.*

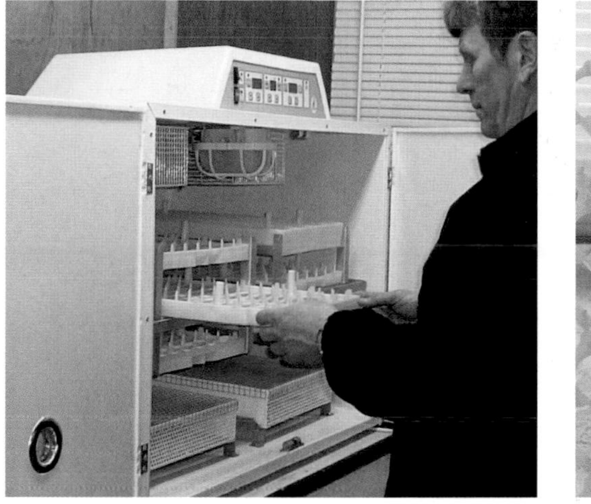

Right *Top-of-the-line incubators have racks for different sized eggs, automatic egg-turning mechanisms, and fan-assisted air temperature control systems.* Far right *All incubator components must be completely clean before use.*

a bit before hatching rates reach acceptable levels—you might expect a 60 percent success rate. If you are a complete beginner, it is probably best to start with a small, manual four- or six-egg unit, which will be cheap to buy and relatively simple to use. These devices are little more than a plastic base unit incorporating a water container and an egg tray. The removable lid is usually made from clear plastic and will normally house the lightbulb and control unit; there will also be a couple of adjustable air vents.

Because these basic machines are so simple, you must take care to locate them sensibly. Do not put them in a hot, centrally heated room or in a cold garage; they are not designed to cope with extremes of ambient temperature. The best place is a spare bedroom, with the radiators switched off. This will ensure that the environment remains temperate and reasonably humid, which will allow the unit to function properly. Never place an incubator in direct sunlight.

Eggs in incubators need to be maintained at a specific temperature for the best and most predictable hatching results. Still-air machines should run at 103°F (39.4°C), while the forced-air units can be slightly cooler, at 100°F (37.7°C). You will obviously need a thermometer to keep an eye on this, and readings should be taken from just above the top of the eggs as they lie in the tray. The more sophisticated machines will have temperature adjustment controls. Remember that variations of just 1°C can have a detrimental effect on hatching.

Carefully read the instructions supplied with your machine and follow the recommendations made about location, vent setting, and so on. It is a good idea to run the machine for 24 hours before using it for real, just to make sure that it gets to and maintains the correct temperature.

Clean machine

Cleanliness is a vital factor when you use an incubator. Even a brand new unit will need to be wiped out with a suitable disinfectant before it is first used, and a machine that has been used before should always be thoroughly cleaned before it is pressed into service again. Be careful with the choice of cleaning product. Some of the more aggressive liquids can damage the electronics within the machine, so it is best to use an incubator disinfectant bought from your local poultry specialist supplier.

Choosing the eggs

Once your incubator is set up and ready to go, select your eggs carefully. They should be as fresh as possible—ideally, never more than a week old when they go into the machine. The best results will come from eggs that are up to seven days old. If they are older than this, the success rate will fall dramatically.

They must be clean, too, to minimize the risk of bacterial contamination. An incubator provides an ideal breeding ground for all sorts of germs and pests. Use a recognized egg wash to clean the shells thoroughly.

Avoid using eggs that show any sort of shell defect or are misshapen, including those that are sausage-shaped or over-round. Do not use any eggs that are too small or too large. Stick to those showing the characteristic blunt and pointed ends and those that weigh at least 2 oz (55 g).

Loading the eggs into the incubator is known as setting, and the way it is done depends on the design of the egg tray in your machine. Follow the manufacturer's instructions carefully.

Above *Follow the incubator instructions carefully, and be prepared for a few failures at first. Note here the difference in size between the white Serama bantam eggs and the one from a large fowl.*

Above *If possible, visit the egg supplier in person and check the stock. The chicks you hatch will only be as good as the birds that laid them.*

Are they fertile?

Eggs will hatch in an incubator only if they are fertile—that is, if they have been produced by hens that have been running with a cockerel. Many poultry enthusiasts in urban areas do not keep cockerels because of the potential noise problems, so fertile eggs for incubation will have to be purchased. Quality and freshness are the all-important factors. The best source is likely to be a local supplier, who has been recommended to you by a breed club or poultry society.

Increasingly, people are selling hatching eggs by mail order or through the Internet. Sending eggs through the mail carries obvious risks, of course, and should things go wrong your rights will be more limited than if you have collected the eggs yourself from a local breeder.

You will be able to tell if an egg is fertile only by checking it after it has been in the incubator for about ten days. It is done by a technique called candling. It is simple and easy to do and involves nothing more than shining a bright light through the egg to highlight what is happening inside. If the egg is fertile and development is progressing normally, you should be able to see a spidery pattern (blood vessels)

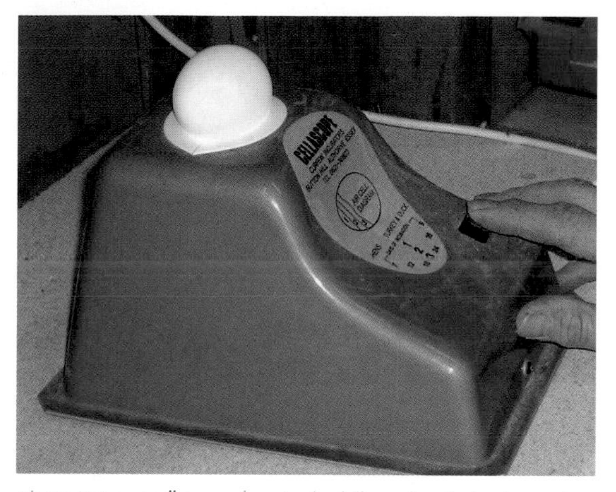

Above *Use a candling machine to check the embryo's development. Eggs with no sign of life should be discarded.*

emanating from a darker, central spot (the embryo). If the egg is infertile or the embryo has died, it will appear completely clear, at which point it should be removed and discarded. You can buy custom-made candling lights, which make the job a lot easier, although they are not essential.

Humidity levels

After temperature, egg-turning, and general cleanliness, the other important factor governing the likelihood of successful incubation is humidity. All incubators include a water tray, but how and when it is filled depends on the design of the machine; refer to the instructions for specific advice.

Humidity levels inside the incubator control the level of fluid that evaporates from within the egg, which is essential to the chick's development. Monitoring this can be done with careful candling to assess the relative size of the air space (found at the wide, blunt end of the egg). This will get progressively larger as the embryo develops and fluid from inside the egg evaporates out through the shell.

The diagram below shows how the air space increases during the incubation period. If, at any point, you feel that the space is looking too small, humidity levels inside the incubator are probably too high and the water should be removed for a day or two. This will increase the rate of evaporation, and the air space will increase in size accordingly. Conversely, if it looks too large, more water should be added.

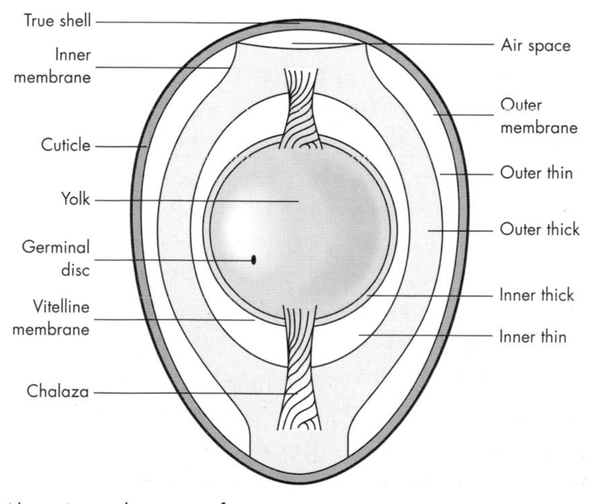

Above *Internal structure of an egg.*

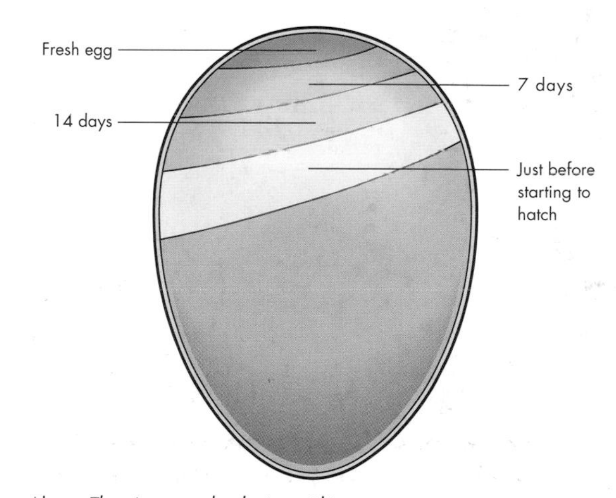

Above *The air space developing within an egg.*

Moment of truth

You'll never get tired of watching chicks hatch, but there's plenty to consider during those first few days of life. Unfortunately, there are bound to be losses, and you'll probably make mistakes, too—always try to learn from what you do wrong.

Hatching

Hatching should occur on day 21. Do not turn the eggs for the final three days of the process, but it is important to keep humidity levels up during this time and just after hatching to prevent the shell membranes from drying out and trapping the chicks.

The chicks will chip their way out of the shell (a process known as pipping), but any that are slow to hatch can be left for an additional 24 hours and no longer. There is usually a reason why a chick doesn't break free, and it is rarely good. Those that fail to hatch should be disposed of.

Leave the newly hatched chicks in the incubator for 24 hours so that they dry out and fluff up fully. Their condition at this stage can provide valuable clues about how well the incubation process has been managed.

Above *This is what it's is all about! Chicks should always be allowed time to dry out before being removed from an incubator.*

Above *Newly hatched chicks are vulnerable and need to be looked after very carefully.*

The young birds should be soft but firm. If the chicks appear dry, it is likely that either the humidity level was too low or the temperature was too high. Bloated, sticky (or drowned) chicks point to high humidity or a temperature that was too low.

Just before a chick hatches, they consume what is left of the yolk and this provides them with the energy reserves needed to tide them over during the first two days of life. The limited nature of these resources mean that it is vital for keepers to create the right living environment for the new arrivals.

Growing up

Once they are hatched and thoroughly dried, the chicks must be moved out of the incubator and into new, larger surroundings. This is a critical stage in their lives, and they must be kept warm and encouraged to eat and drink as soon as possible. In the absence of a mother hen who will lead by example, the responsibility falls on the keeper to provide the necessary support.

Brooding is the term used to describe the first month or so of a chick's life, after which he or she passes into the rearing or growing stage. This continues right through until the 18-week mark, when hens reach the point-of-lay (POL) and start producing eggs of their own.

The brooding period is very important, because the chicks are vulnerable to environmental conditions and diseases. For this reason, they need to spend these early weeks in a broody enclosure, which will protect them from draughts. The enclosure must have an adequate heat source to keep the chicks warm, and the size will obviously depend on the number of chicks you are dealing with. If there is only a handful, you can get away with using a large cardboard box with a 60-watt lightbulb suspended above it. A better approach is to set aside a shed for the purpose, and this is what experienced breeders do. Make sure that it is scrupulously cleaned and disinfected after each batch of chicks has passed through. Brooder houses need to be weatherproof and rodent-proof, and they should not contain any other poultry at the same time.

The chicks will need litter on the floor—clean white wood shavings are the best choice—spread to a depth of 2–3 inches (5–8 cm). At first the young birds will need to be enclosed in a small area bounded by walls of cardboard or hardboard, which needn't be any higher than about 18 inches (45 cm). It is also a good idea to place an old towel or sheets of corrugated cardboard on top of the litter to help the chicks find their feet. The firmer surface will make it easier for them to stand and move about initially, and it can be

Above *Handling chicks from an early age helps them get used to human contact.*

removed when they look confident with their movements. The movable walls will stop the young chicks from straying too far from the heat source.

Vital warmth

Temperature control is a critical factor in the early weeks while the chicks are growing their first set of feathers. The heat source, or brooder, should be placed in the center of the enclosure. It can be a free-standing unit or one that is suspended from above. Some of the most popular are similar to infrared heat lamps, emitting either red or white light, but the best sort emit no light at all, just heat. These are called dull emitters.

The brooders that produce white light are probably the least desirable. Too much light can cause the young birds to develop too quickly, and they will be nervous and unwilling to settle at night. It may also promote undesirable behavior such as feather pecking. Moreover, the blood that this inevitably produces acts as a stimulant, worsening the whole situation. This is why some keepers prefer heaters that emit red light, because any blood appears black, rather than red, but this really is poor consolation. In practice, it is best to have your light and heating sources completely separate, so that both can be controlled independently.

Left *It is vital to keep young chicks warm as their first set of feathers are growing.*

Left *The fact that these chicks seem to be bunched together under the heat lamp indicates that they might be too cold.*

Getting the temperature right is vital. You should seek advice about the best setup for your own situation from the brooder supplier or from an experienced breeder. In general, however, the young birds will need most heat at the start of the brooding period and the least at the end. Therefore, the height at which the heat source is suspended will need to be varied, depending on the size and breed of the birds involved. Use a height of 16–18 inches (40–45 cm) to begin with. Mounting the unit on a chain, so that the height can be adjusted link by link, makes life much easier. It will have to be raised a little each week for the first five or six weeks, after which the first set of feathers should be in place, and the birds will be better able to regulate their own temperature.

Watch and learn

There are so many potential variables involved in hatching and rearing chicks that it is difficult to give specific advice. However, much can be learned from the behavior of the chicks. If, for example, you find them all huddled under the brooder with little sign of activity, they are cold and the brooder should be lowered. On the other hand, if they are panting and scattered all around the perimeter of the enclosure, the heat source is too low and providing too much heat. Finding all the birds grouped together in one corner can indicate that there is a draught in the enclosure that should be blocked off.

Contented and comfortable chicks will be spread out all over the enclosure and showing various degrees of animated behavior. They will also be making short, "chip, chip" noises. Long-drawn-out cheeping is the sign of an unhappy chick.

The ambient temperature of the brooder room or shed plays an important part in the process, too, and keepers should not rely on the brooder as the heat source for the whole area. As a guide, chicks in their first week of life need a room temperature of at least 66–77°F (19–25°C), after which it can be progressively reduced to about 60–61°F (16°C) by the end of the first month. Chicks in the first few weeks of life are very susceptible to low ambient temperatures. If they are too cold, they will find it difficult to eat and drink and will become dehydrated and die. For this reason, it is important to get the room thoroughly warmed up before the chicks are installed—turn the heating on the day before to make sure. If you want to be doubly sure during the setup procedure, you can place a thermometer on the shavings, directly under the center of the brooder. For the best results, the temperature recorded here should be around 93°F (34°C).

Droppings sticking to the chicks' abdomens are a sign that they are suffering from a chill or draughts. Remove the droppings carefully, cover the area with a dab of petroleum jelly, and check the temperature.

Food for thought

To encourage the young birds to eat you must provide plenty of the right sort of food. A starter crumb for chicks, which is perfectly balanced and sized for very young birds, is ideal. This can be placed in dispensers around the perimeter of the brooding enclosure, but it must be readily available.

All feeders should be kept full for the first eight weeks or so. You can encourage and teach chicks that do not have a broody to eat by including a little finely chopped hard-boiled egg or spring onion with the food.

You should also make a little hard grit (granite and flint) available to the birds once they are a week old, sprinkling it lightly on their food. Make sure that it is the correct size for chicks. It will aid the development of their digestive system and gizzard. Add a little grit every two weeks, and then switch to a slightly larger dose of growers grit at the six-week point.

Water is possibly even more important than food. A chick can lose all of his fat and half his body weight and still survive. However, if just one-tenth of his body's moisture content is lost he will die. It is vital, therefore, to have water available so that the chicks can start drinking as soon as they get into the brooder pen. Use shallow drinkers (to reduce the risk of accidental drowning) and place them nearer the center than the feeders, although not under the heat source. Change the water and clean the containers every day to prevent the buildup of slime.

Good lighting

Keeping the brooder pen well lit during the day will encourage the chicks to feed and drink, but bear in mind that too much light can over-stimulate them. At the other extreme, when the lighting is too dim, they will not eat and drink properly, which will affect their overall growth rates and healthy development.

If you have set up the brooder pen in a bright shed with large windows, curtains can be used to limit the light, but take care that you do not inhibit the shed's ventilation, as this is another important requirement of a brooder house. Some people keep them too hot, believing that the birds will prefer higher temperatures. They do not, and such conditions can be harmful.

Great growers

To grow healthily at this early stage, the young birds need enough space to be able to move around, particularly at night. If you were a fly on the wall in the brooder house once the lights were turned off, you would notice an almost constant rotation among the birds as they swap places under the brooder. They need to have the space to move out and away from the heat source when they get too hot. It is important that this is allowed to happen to encourage the chicks to develop and to promote good feather growth. Birds that are kept in a cardboard box for too long will grow unevenly, will show poor feathering, and may even turn out to be more susceptible to disease later on, as well as other weakening conditions.

By the time the birds get to be six or seven weeks old, you will need to start thinking about switching off the brooder and moving them outside. Do this gradually to avoid any sudden changes that will promote stress. Much will depend on the weather and the time of year, of course, but assuming your birds have reached this sort of age in spring, give them a couple more weeks inside the brooder shed, but with the heat turned off to toughen them up.

At the eight-week mark you should also switch them to growers pellets or mash, phasing in the change over about two weeks by gradually adding more of the growers feed. Do this before you turn them out and make sure that they are used to the new food. Even something as apparently trivial as a change in the feed type will induce a degree of stress for the birds—they thrive on routine, remember. Combining a feed change with a move to a henhouse outdoors would be a major, and unnecessary, trauma for them to deal with. Get them happy and settled with the new food, then move them into a henhouse and run outside.

Keep them away from any other poultry you may have because they will still be vulnerable to infection. They should go on to fresh, clean, short grass, well away from other stock. Do not put them on grass that has been recently cut and is covered with clippings. These can be indigestible for young birds and will cause compaction in the crop or gizzard, which can result in death.

Finally, it is good practice to always deal with your youngest birds first when it comes to daily feeding and watering. The last thing you want to do is carry droppings from older birds into your growers pen on the bottom of your boots—it is easily done and will quickly spread disease. If you go to the young birds first, with clean boots, you're unlikely to have a problem. This is a simple common sense measure, but one that is well worth remembering.

Above *Moving young birds from the brooder to an outside henhouse is an important stage—the timing must be right.*

Showing

Exhibiting poultry is an important part of the hobby. Even though it is taken seriously by only a minority of keepers, it is at the core of maintaining the breeds as we know them.

Exhibiting poultry

Showing can become addictive, and you can find yourself starting to take your birds' condition increasingly seriously. Breeding is the key to success, and it is important to appreciate that producing a winner won't happen overnight.

National standards

To be successful at a top show, a bird must conform perfectly to the standard for his breed—he must have all the right bits in all the right places. The breed standards are a well-established set of guidelines that lay down how each breed should look under ideal circumstances. For example, plumage color, comb type, feather type, and overall weight will all be specifically detailed, and birds entered into a show will be judged on the basis of the rules. National standards are set and published by poultry clubs and associations, and these give details of the fundamental characteristics of every established breed, as well as listing the typical defects to be avoided. In the US this is undertaken by the American Poultry Association and the American Bantam Association; in Britain it is done by the Poultry Club of Great Britain and the Rare Poultry Society.

Above *Trophies and ribbons are the main rewards for success in poultry showing. Cash prizes are generally small.*

The best breeders tend to be the experienced ones. They understand the requirements for their chosen breed and have a practical understanding of poultry genetics. This is a complicated subject, and it takes most keepers years of carefully monitored breeding before they gain a useful working knowledge about what affects what.

Choosing a breed

This is not to say that beginners cannot get involved in showing. Everyone has to start somewhere, and most poultry clubs and regional societies organize a number of informal shows each year. These are a great way for novices to get an introduction to exhibiting, and beginners are always welcome. Events such as these provide excellent opportunities for those interested in getting started to see the range of birds available and to meet experienced breeders.

Most of the major breed clubs will be represented at these events, and joining the one that supports the breed in which you are interested is always a sensible first step. These organizations brim with enthusiasts who will be happy to help anyone who is just getting started, and they will know where to buy the best birds.

Of course, with 100 or so breeds to choose from, deciding on which to keep for exhibition purposes can be difficult. The sensible advice is to take one of the easier options to begin with. Breeds like Polands and Sultans are classed as high maintenance because of the birds' crests and feathered legs. These are not ideal for the beginner, and novices would be far better advised to pick a simpler breed, such as the Welsummer or Rhode Island Red. Similarly, Orpingtons, Wyandottes, New Hampshire Reds, and Light Sussex can make great starter birds for newcomers to showing. Remember, too, that if you are hoping for success you will have to buy quality show stock. The chances of a novice breeding birds that are good enough from scratch are remote.

THE TYPE

If you are planning to show your chickens, you need to understand what judges are looking for when they assess each bird. The type is just as important as the condition of the feathers.

– **Light breed** These birds are best known for their egg-laying abilities. Many originate from the Mediterranean region and produce plenty of light or white-shelled eggs. They are not regarded as good table birds, and often have "flighty" characters, making them excitable and nervous. Typical examples include Ancona, Minorca, and Spanish.

– **Heavy breed** This group consists of breeds developed with utility production in mind. They are good for the table and can be acceptable layers, too. They tend to be large and feathery and have docile, gentle natures, which make them ideally suited to the domestic environment. They are great first birds. Typical examples include Rhode Island Red, Orpington, and Sussex.

– **Hard feather** These birds have tight, short feathering, which closely follows the shape of the body. The term is applied to game birds, which were historically used for cock fighting. Typical examples include Malay, Modern Game, and Old English Game.

– **Soft feather** The classification encompasses all breeds that are not hard-feathered. The feathering is generally loose and fluffy and will often disguise the bird's body shape. Typical examples include Brahma, Plymouth Rock, and Silkie.

– **Bantam** The miniature version of a conventional chicken breed is often identical to the large fowl relation. Bantams are now accepted as being one-quarter the weight of the parent breed. Bantam versions of almost all popular chicken breeds are now available.

– **True bantam** Unlike ordinary bantams, a true bantam is a naturally small bird that has no large fowl counterpart. It is a breed in its own right. Most true bantams have no utility role and are kept purely for their beauty and for exhibition purposes. Typical examples include Pekin, Rosecomb, and Sebright.

– **Rare breed** These are breeds that do not have their own specialist club or association to support their interests and survival. Typical examples include Andalusian, Houdan, and Norfolk Grey.

Training a winner

You should not expect to be able to put a bird straight into a show without spending a great deal of time and effort cosseting and training him first. Preparation is the key—a good bird requires careful conditioning.

Pampered existence

It is worth bearing in mind that any birds you hope to show seriously will require a more pampered existence than most domestic birds enjoy. For a start, they will need to spend their time under cover. Direct sunshine will affect plumage color, fading darker feathers and giving white ones a pale yellow tinge. Allowing potential show chickens to free-range with other birds will not only make them much harder to prepare—they will get dirty outside—but may also lead to unnecessary feather damage.

Secrets of success

Chickens are creatures of habit. They love routine and are easily upset by changes in their way of life. So birds that are suddenly thrust into an exhibition environment become understandably upset, which is why they need to be trained for the experience. Most domestic poultry will not be used to the relatively confined environment of a show pen, so it is important that you get them acclimatized to this aspect before the event. In addition, the bird will need to be relaxed about being handled, because the judge removes each bird from his pen during the assessment process. The last thing the judge will want is to be confronted by a bird that flaps wildly during the inspection. An irritated judge does not tend to award top marks!

Regular confinement and handling during the weeks leading up to the event should do the trick. Of course, the breed of bird plays an important part. Those breeds

Above *Birds must be used to being handled if they are to come through the judging process successfully.*

Above *Use an old brush to clean the bird's legs and feet. This is an aspect often ignored by beginners.*

Above *Great emphasis is placed on a bird's head. The comb, wattles, beak, and eyes must be perfect.*

that are known for being active and flighty, such as Hamburghs, Leghorns, or Minorcas, will be much more of a handful in this respect.

Once you have chosen the event you want to enter, you will need to apply for what's called the show schedule. This will include the entry form, plus all the additional information about timings and so on. For example, birds need to be penned-up by a certain time so that they are ready for judging, and there are also usually restrictions on when birds can be removed after the event. Entry fees are normally low, as, in most cases, is the prize money. Showing is certainly not a get-rich-quick exercise.

COMB TYPES

The comb is the fleshy protuberance on top of a chicken's head. Combs are normally larger on male birds, and there are many different forms. The main types of comb are:

- **Cup** As the name suggests, this type of comb looks rather like a small cup with a serrated edge.
- **Horn** This type normally looks more like a V than actual horns. It begins just above the beak, then branches into two, tapering spikes.
- **Leaf** The comb looks like an open-winged butterfly on the front of the bird's head.
- **Pea** This triple comb looks like three small, single combs arranged side by side. The one in the center is normally slightly larger than the other two.
- **Raspberry** The comb resembles half a raspberry, its surface being covered in small, rounded nobbles.
- **Rose** A broad, solid comb, which is almost flat on top and covered with small points. This type also features a rearward-facing spike known, oddly, as the leader.
- **Single** A narrow, usually upright blade with serrations of various depths along the top edge. It varies in size from breed to breed.
- **Strawberry** This type resembles the outside of a strawberry; it is also sometimes referred to as the half-walnut comb.

Good prepartion

As with all livestock showing, one of the primary secrets of success when exhibiting a chicken is good preparation, and there really is no excuse for getting this wrong. Even though you may not have a bird that is technically perfect, you can certainly make sure that he is clean and tidy for the judges.

It is expected that all entrants will have been washed and dried before the event. This must be done by hand and with great care—breaking just one important feather could seriously damage your chicken's chances.

Stand the bird in a bowl or sink (whichever is best suited to his size) containing about 3 inches (8 cm) of warm water. Take things slowly and gently so that you do not startle the bird. Then, once he is settled, start to wet the bird all over, taking care not to splash too

Above *Poultry showing is about prestige and satisfaction, not money-making.*

much on the face. Add some shampoo and gently massage the feathers. Only use a mild product, such as a baby shampoo; never use anything too aggressive. You can buy special, anti-mite poultry shampoos, and these are ideal.

Work from behind the bird's head toward the tail, keeping one hand on the bird's back to keep him settled while you wash. Wash away from the head, not against the grain of the feathers. Take great care to avoid breaking any of the feathers as you work in the shampoo, and pay special attention to the vent area, the legs, and the feet. Use an old toothbrush or nailbrush to clean the scales on the legs and feet, and keep adding fresh, warm water as you work. Birds can become chilled very easily.

Thoroughly rinse off the shampoo in a separate bowl or sink, again with warm water. Careful drying is essential, and your approach will probably be governed by the time of year. If the weather is warm, the bird will be happy to dry himself outside in a sunny run—remember to provide plenty of clean litter, of course. If it is hot, make sure that there is some shade in the run so that the chicken is not in intense summer sun.

In cooler weather you will have to do the drying by hand, using clean towels and a hair dryer. Remove most of the excess water by hand—again take care that you do not damage the feathers—and then use a hair dryer for the final finishing. Never use it on its hottest setting, and keep the nozzle moving at all times—it is important to avoid localized overheating, which is obviously uncomfortable for the bird and can also damage the feathers.

As a rule, washing is best done three or four days before the show. This will allow time for the natural oils to be returned to the feathers, ensuring that they regain that all-important, natural-looking sheen.

Finishing touches

Once the main wash and dry are completed, you are left with the final detailing. The bird's comb and wattles need to be cleaned, then dressed to intensify the color and produce a lustrous finish. Experienced exhibitors have their own special ways of doing this, but they tend to keep them secret—which underlines the importance of this stage. The head is the focal point of any bird, so being able to give the bright red comb and wattles an extra special finish can count for a lot on the show bench.

Unless you can wheedle some useful information out of someone who knows better, satisfy yourself with some gentle cleaning with a soft cloth or sponge and mild soapy water. Rinse off carefully, dry and then treat the areas to a smear of a product based on petroleum jelly. Alternatively, you can create a similar effect by using vegetable oil.

At the show

One of the few drawbacks to the whole business of poultry showing is that the birds that are being exhibited can be exposed to both diseases and parasites. Sensible exhibitors keep their show bird isolated from the rest of their flock after an event.

Damage limitation

The fact that a large number of birds are gathered together in one room, with the chickens contained in pens that are packed tightly, side-by-side, means that contagious conditions like respiratory problems and lice or mite infestations can spread easily.

The onus is always on owners to make sure that their individual birds are clean and healthy before they attend a show. However, people's vigilance varies in this respect, and some owners are a good deal more particular than others. Although judges have the power to refuse to assess a bird that is obviously ill or infested with mites, it remains quite common for birds in this sort of condition to be presented for showing.

It makes a lot of sense to dust your birds with an anti-parasite powder as soon as you get them home. Experienced exhibitors are also careful to keep their show birds separate from the main flock specifically to avoid the risk of needless cross-contamination.

On the move

Finally, a word about transporting your poultry to and from a show—or anywhere else for that matter. This is a potentially stressful business for both you and the birds. Chickens will find it unsettling no matter how you do it, so do your best to minimize their discomfort.

It is possible to buy boxes and cages that have been specifically designed for containing poultry, but they are expensive. As a result, most keepers prefer to use cardboard boxes instead. These are perfectly acceptable as long as they are large enough (but not too large), adequately ventilated, and secure. The bird must be able to stand comfortably inside the box with the lid closed, and three or four holes should be made at the top of each side to ensure good air circulation.

Do not be tempted to use an enormous box in the belief that the bird will be happier. In fact, too much space will cause the chicken to slide around, which will be distressing for the bird and may harm him. Line the box with newspaper and cover this with straw or wood shavings. Never forget to secure the lid of the box.

Make sure that the boxes are wedged or supported in the car to prevent them from sliding all over the place at the first corner. Your driving style should reflect the delicate cargo you are carrying, and you must remember to keep the car well ventilated at all times, both while you are driving and while you are parked. Chickens do not cope well with excessive heat, so if possible avoid attempting to move them on hot days. If you have to drive a long distance, you might consider stopping on occasion to give the birds a drink.

Left *Birds need to be settled and relaxed in the judge's hands.*

Chicken Breeds

This section showcases more than 50 of the most interesting and colorful poultry breeds. To make things simpler for you, they've been divided into four main categories—hard feather, soft feather, true bantam, and rare breed. These are the established classifications used in the domestic poultry world.

Indian Game

striking looks • muscular • poor layer • exotic breed

TYPE: large, heavy, hard feather • WEIGHT: large male 8 lb (3.6 kg), large female 6 lb (2.7 kg); bantam male 4 1/2 lb (2 kg), bantam female 3 1/2 lb (1.5 kg) • COLORS: Dark, Double-Laced Blue, Jubilee

Despite the exotic-sounding name, this breed was actually created in Cornwall, in southwest England, in the 19th century. Enthusiastic breeders are thought to have crossed strains of Malay, Asil, and Old English Game birds to produce what we have today. In America, the breed is known as the Cornish Game, which seems reasonable, although it does not reflect the breed's Far Eastern roots.

The Indian Game is widely recognized as being one of the best table birds there is, and for this reason he has been used over the years to cross with other established table birds, such as the Dorking and Sussex, to produce excellent, large offspring. American breeders are believed to have used these birds in the development of some of the early broiler birds.

Looks

The Indian Game is a striking-looking bird. He has a purposeful, muscular body, hard feathering, and domineering stance. The wide-breasted body is particularly broad at the shoulder. The medium-sized head has bold eyes and a stubby, powerful beak. The comb, if left undubbed, is pea-type, with three ridges running from front to back (the central one being the highest). The bird stands on strong, thick, widely spaced, orange-yellow legs, which dominate his appearance. His four-toed feet are large and well spread.

The male Dark Indian Game is essentially black, although much of the feathering has fine bay or chestnut detailing. The female has a dark brown ground color, but plenty of black, too, with the same beetle-green sheen as the male.

The male Jubilee is predominantly white, with bay or dark brown trimmings, most noticeably on the neck and wings. Again, the female is much darker than the male, but with a light head and neck and attractive white lacing in many areas, including the breast.

The male Double-Laced Blue has a two-tone color, with dark blue and grey on top, lighter beneath, and dark brown edging here and there. The female bird has dark grey and blue on top, but is dark brown everywhere else, with light blue lacing.

Personality

This breed is sometimes docile and friendly, but sometimes aggressive—even the females.

Eggs

Not a breed known for its laying ability, the Indian Game will typically produce only about 80 small, light brown eggs a year.

Day-to-day

The Indian Game's strength makes these rugged, robust birds, although they are not particularly active. They will cope well with low temperatures and are easy to contain. The hens can make attentive mothers, but they may be aggressive. In addition, some strains exhibit hatching problems. Fertility can be a major problem because of the width of the birds—they find it very difficult to mate properly. This is not a breed that is ideally suited to inexperienced keepers.

Dark Indian Game bantam male

Modern Game

unique looks • exhibition favorite • hardy • sometimes noisy

TYPE: large, hard feather • WEIGHT: large male 7–9 lb (3.2–4.1 kg), large female 5–7 lb (2.25–3.2 kg); bantam male 20–22 oz (570–620 g), bantam female 16–18 oz (450–510 g) • COLORS: Black, Black-Red, Blue, Blue-Red, Birchen, Brown-Red, Golden Duckwing, Lemon-Blue, Pile, Silver-Blue, Silver Duckwing, Wheaten, White

When cock fighting was finally outlawed in Britain in 1849, many breeders who still sought an element of competition with their birds found refuge in the burgeoning poultry exhibition scene. These shows grew dramatically in popularity during the 1850s, and with this came the desire to produce bigger and more feathery exhibition birds.

The game fanciers could do little about the hard feathering on their birds, which was traditionally short and close fitting, but they could certainly increase the overall size of the birds. The most common breeding cross used to achieve this was with the Malay, a bird that introduced a good deal more height and brought in different tail characteristics. These were elegant, specialized creatures, available in a limited range of colors and developed specifically for exhibition. But it was not until the formation of the Old English Game Club that a distinction was finally made and they became classified as Modern Game.

The large fowl version had a relatively brief era of popularity, which reached a height in the early 1900s. But the birds' status started to decline, and by the end of the Second World War the type remained in the hands of just a few dedicated enthusiasts.

The bantam version, however, continues to be treasured among both keepers interested in exhibition and those who simply appreciate the breed for its classically simple, attractive lines. Color, type, and style remain the most important factors from a judging point of view, and there are currently 13 standardized versions available.

Looks

The Modern Game has a look all its own. The body is short with a flat back, and the whole thing should resemble the shape of an iron when viewed from above—tapering toward the tail. Both the wings and the tail should be short; males have narrow, pointed, and slightly curved sickle feathers.

The head is long and narrow with prominent eyes, which vary from red to black, a small single comb, wattles, and earlobes, and a relatively long beak. The neck is long, slightly curved, and covered with wiry feathering.

Standing on long legs, the Modern Game displays muscular thighs and rounded, featherless shanks. Each foot has four toes, and the leg and foot coloration includes yellow, willow green, and black, according to overall coloring. The combs and wattles

Birchen Modern Game bantam female

can vary in color, too, ranging from red to dark purple and black, depending on feather color.

Personality
Opinions vary on the Modern Game's personality. Some keepers describe them as pleasant, friendly, easy-to-handle birds, but others suggest that they are aggressive and noisy. The truth probably lies somewhere between, with a lot depending on how well the birds have been reared.

Eggs
This is not one of the world's great laying breeds. You should expect about 90 small, light-colored eggs in a good season.

Day-to-day
The Modern Game is a hardy breed that loves to be active if given the space to do so. These birds are not generally as aggressive as other game breeds, so they can be kept with other breeds under careful supervision. The female birds do go broody and can make good, protective mothers, although their hard, short feathering and long legs do not really provide the best brooding combination.

Black-Red Modern Game female

Old English Game

fantastic coloring • striking looks • noisy • aggression in the large fowl

TYPE: large, hard feather • WEIGHT: large male 4–5 lb (1.8–2.55 kg), large female 2–5 lb (0.9–2.55 kg); bantam male 22–26 oz (620–740 g), bantam female 18–22 oz (510–620 g) • COLORS: an extremely wide range, including Black-Red, Blue, Brown-Red, Crele, Cuckoo, Duckwing, Partridge, Splash

As its name suggests, the Old English Game is an ancient breed that was even documented by the Romans. Julius Caesar reported that the Britons kept fowl for "pleasure and diversion," and it is likely that the word "diversion" referred to cock fighting. This cruel pastime was widely practiced by all sorts of owners, from peasants to princes. The male birds were set against each other, often in specially made pits and wearing metal spurs on their legs. Cock fighting was banned in England in 1849, although the practice still goes on today as an underground activity, however, not to the extent it used to.

Most people are now interested in showing. The Old English Game Club was founded in 1887, and since then two primary types have evolved: the Oxford and the Carlisle.

Looks

The Carlisle type is broad-shouldered with a full breast and a bold but graceful appearance. The head is small and features what is known as a "boxing" beak—that is, the upper mandible closes tightly over the lower one, somewhat like a hawk's beak. The comb, wattles, and earlobes are all small and fine. The eyes are large and the head sits on a long neck that is thick, particularly at the base. The legs are strong and featherless.

The Oxford type shares many of the same characteristics, although he is more upright and certainly appears more confrontational.

Personality

The breed's fighting origins mean that these birds tend to be aggressive and self-contained. They are not suited to mixed flocks.

Eggs

The birds are reasonable layers of small, tinted eggs, and owners can expect about 120 in a season.

Day-to-day

This breed is not ideal for a beginner. These birds need space, love to forage, and tend to be noisy. They are hardy and do not like being confined. The hens make protective mothers. Males can sometimes adopt the brood and assist in rearing the chicks. This habit is virtually unseen in any other poultry breed.

Splash Old English Game bantam female

Black-Red Old English Game bantam male

Rumpless Game

unusual looks • plenty of character • quick-tempered • can be noisy

TYPE: large, hard feather • WEIGHT: large male 5–6 lb (2.25–2.7 kg), large female 4–5 lb (1.8–2.25 kg); bantam male 22–26 oz (620–740 g), bantam female 18–22 oz (510–620 g) • COLORS: various

The Rumpless Game is certainly an odd-looking bird—the sort of chicken you either love or hate. Many people find the lack of a tail bizarre and unnatural, but other keepers believe that the birds have lots of charm and character.

The bird has no tail because he does not have a parson's nose, the part of the body technically (but less quaintly) known as the caudal appendage, the fleshy protuberance from which the tail feathers grow on a normal bird. Its absence appears to have been little more than a genetic fluke, and the condition has been known for centuries.

The precise origins of this bird are not clear, although there are records of fowl without tails existing in various places around the world, including India, China, America (Virginia), and the West Indies. There are other examples of the occurrence of similar characteristics like this, including the frizzle-feathered fowl of Mauritius and the tailless cats from the Isle of Man.

Although it is possible for Rumpless Game to exist in a large fowl form, they are far more common as bantams.

Looks

These birds have a bold, upright carriage and are very similar to traditional game birds, apart from the complete lack of a tail. They are small and have a steeply sloping profile, with a full and prominent breast and large wings. The head is small, with quite a large beak, bold eyes and a thin, single comb. Earlobes and wattles are small. The feathering all over is hard and close. These birds stand on strong legs, with four toes on each foot.

The breed's plumage coloring is variable. In competition, color is not an important factor; more emphasis is placed on the bird's head and overall appearance.

Personality

The breed shares typical game characteristics when it comes to personality, so they can be peppery little birds. However, many keepers say that they can be easily tamed and that they make excellent pets.

Eggs

For their size, Rumpless Game hens can be surprisingly good layers of light-colored eggs.

Day-to-day

In their bantam form, these birds are happy to be confined, although they can be noisy. They are hardy, too, although some owners do experience problems with the fertility of eggs. It has been suggested that the feathering around the hen's rump plays a part in this and that its removal from the vent area can improve matters.

Above *Cuckoo Rumpless Game bantam male*
Right *Wheaten Rumpless Game bantam female*

Australorp

great all-rounder • docile and quiet • good mother • happy being handled

TYPE: large, heavy, soft feather • WEIGHT: large male 8 1/2–10 lb (3.85–4.55 kg), large female 6 1/2–8 lb (2.95–3.6 kg); bantam male 36 oz (1.02 kg), bantam female 28 oz (790 g) • COLORS: Black, Blue

The Australorp is a breed that has much to offer the domestic keeper, but which is frequently overlooked by those who decide to opt for one of the more mainstream breeds.

The breed was developed in Australia using, among others, Black Orpington stock that had been taken there from Britain, and the name represents a combination of Australian and Orpington. Great care was taken during the breed's development to avoid sacrificing overall size while retaining good laying capabilities.

The Australorp's popularity as a good-looking, dual-purpose bird has since spread around the world, and now he is universally—and rightly—regarded as a great all-round performer.

Looks

The black plumage has a stunning beetle-green sheen when it catches the sunshine, and this contrasts wonderfully with the bright red single comb, face, and wattles.

The birds generally adopt an upright stance, giving them an active and graceful appearance. Their bodies are deep and broad,

Black Australorp male

especially across the shoulders and saddle area. The tail is full yet compact, and the birds stand on strong, dark, featherless legs. The beak, too, is dark, as are the eyes. Just about the only areas that are not either black or red are the soles of the feet, which are white.

The Blue variety is subtly different from the Black type, showing a combination of dark and medium slate blue, and there is some lacing at the ends of the feathers. The legs and beak can be slate blue, too, and ideally the toenails (as well as the soles of the feet) will be white.

Personality

The Australorp's character has plenty to recommend it. Overall, it is a docile, quiet breed. These birds are perfectly happy being handled and so are equally at home with domestic keepers and in the exhibition pen. Their calm temperament also means that the Australorp has great qualities as a broody and an attentive mother.

Eggs

There are stories of fantastic laying records achieved in Australia during the breed's development. One hen reportedly produced 364 eggs in 365 days. On a more practical level, young hens should happily produce 200 eggs in a season, which is a great return. The eggs themselves will be of medium size and light brown in color.

This fine Australorp male has a bright red single comb and glossy black plumage.

Day-to-day

The dual-purpose nature of the Australorp means that as well as producing useful numbers of eggs, they are also good table birds, white-skinned, and generally well fleshed and tasty.

The breed is easy to keep. Their placid character means that the birds tolerate confinement well, but they will happily forage given the opportunity. In addition, they are economical eaters and mature relatively early; however, female birds can become overweight if they are not kept busy.

The Australorp is not generally known as a flier, so there is no need to build an especially secure run to keep them contained. Finally, their hardiness is impressive—they show good resistance to low temperatures and are generally long lived.

Barnevelder

beautiful brown eggs • calm and friendly • good nature • robust

TYPE: large, soft feather • WEIGHT: male: 7–8 lb (3.2–3.6 kg), female: 6–7 lb (2.7–3.2 kg) • COLORS: Black, Double-Laced, Double-Laced Blue, White

The Barnevelder is a comparatively new breed that emerged in the early 1920s. It takes its name from the small town of Barneveld in the Netherlands, and it is believed to have resulted from the crossing of local fowl with imported Asiatic breeds, including the Brahma, the Langshan, and the Malay. The aim of this breeding was to improve the egg-laying performance of the local hens, and this is exactly what happened. As bonuses, the Barnevelder's eggs were large and dark brown, and winter-laying performance was enhanced.

The breed was a great commercial success until competition from the American hybrid layers arose in the 1950s, and the Barnevelder's popularity started to fall away. Nevertheless, it remains a much admired breed among hobby keepers in the Netherlands and the rest of Europe, both for its good looks and its wonderful eggs. There are bantam versions of all the color variants.

Looks

The Barnevelder is characterized by a broad, deep and full body, with a medium-sized head and red face, which is unfeathered. The single comb, wattles, and earlobes should all be bright red, and the eyes should be orange-brown. The bird stands on clean, yellow legs.

The most popular version, the Double-Laced, has dark hackle

Double-Laced Barnevelder female

A group of Barnevelder pullets will thrive in your backyard.

feathers (with a beetle-green sheen), with contrasting edging. The breast is light brown, and there is an attractive brown and black lacing effect on the wings. The male's tail feathers are black, but on the hen the lacing continues up into the tail. On the Double-Laced Blue, the black lacing is replaced by an attractive light blue-gray. The Black and White types are both solid-color versions of the breed.

Personality

This is a great bird for the backyard. It is a calm, docile breed that is ideal for the family environment. The birds are friendly, although their good nature does mean that young birds can suffer from bullying, so be careful if you are keeping them in a mixed flock; it is particularly important to take special care when introducing new birds to the flock.

Eggs

Unfortunately, the production of the dark brown eggs comes at the cost of numbers. Selective breeding for this desirable shell color over the years has affected overall levels of productivity, and the poorer layers produce the best shell color. Nevertheless, keepers should expect about 170 eggs a year from a healthy young hen. Generally, the color of the eggs will gradually lighten as the hen ages. The Double-Laced varieties are regarded as being the best layers of all.

Day-to-day

Barnevelders are easy to live with because the birds are robust and will stay generally healthy, as long as they are kept in good conditions. They eat well and are good foragers, and they do not react badly to confinement. Their clean legs mean that any problems with parasites are easy to spot.

Barnevelders are quick growers, although the cockerels are slow to feather, but this can be a useful identification point for sexing purposes early on. In addition, any day-old Double-Laced males will be easy to identify from their white breasts; the females are gray-brown there.

Brahma

gentle giants • feathered legs • easy to handle • reasonable layer

TYPE: large, heavy, soft feather • WEIGHT: large male 10–12 lb (4.55–5.45 kg), large female 7–9 lb (3.2–4.1 kg); bantam male 38 oz (1.08 kg), bantam female 32 oz (910 g) • COLORS: Buff Columbian, Dark, Gold, Light, White

If you like big, impressive chickens, the Brahma could well be your ideal breed—chickens don't come much larger than this.

Although it is widely listed as a breed from Asia, it is now generally believed that the Brahma was created in America in the 1840s. This exotic-looking bird is thought to have been the result of crosses made from breeds that had been imported to the United States from India (the Grey Chittagong) and China (the Shanghai). Some people have suggested that the Cochin was involved in the mix, too.

Brahma males can feature either a triple or a pea comb.

The first examples of the new breed arrived in Britain in 1852, and they came to prominence largely because of the publicity-seeking efforts of an American breeder, George Burnham, who sent nine birds directly to Queen Victoria and made sure that the papers heard all about it.

The breed name Brahma is derived from an abbreviation of Brahmaputra, one of the great rivers of Asia, which flows from China, through India and then Bangladesh, and out into the Bay of Bengal. The name reflects the breed's Asiatic roots.

Looks

The Brahma—particularly the male—is a striking-looking bird, which manages to be both sedate and active at the same time. He is broad, square, and deep, with a short back, profuse feathering, and large, strongly feathered legs. The small head and short beak seem to accentuate the impression of overall size. The eyes are large, but the comb is small and a triple type.

Long, bright red earlobes flank a featherless face and lead to small, bright red wattles.

The five standard plumage colors noted above are all attractive. Bantam versions are available too, and these are exact replicas of the larger birds.

Personality

The generally gentle nature of the Brahmas and their fine appearance make them popular domestic birds. They are usually friendly and easy to tame, and as such do not mind being handled, although some owners are reported to have found them to be a little standoffish. But do not let this deter you from keeping these gentle giants.

Eggs

The Brahma is a fair layer of medium-sized brown eggs. Unfortunately, laying performance is not what it once was because breeding for plumage color and quality has resulted in the breed losing some of its original egg-laying ability. You should expect about 140 eggs a year from a young, healthy hen.

Day-to-day

Despite his size, a male Brahma can be a timid creature, which makes him susceptible to bullying by other cockerels in mixed, free-range flocks. It is important to watch out for this if you keep your birds on this basis.

They are slow-developing birds, taking up to two years to become fully mature. Another consequence of this slow development and their relaxed attitude can be a tendency to become overweight if overfed.

Generally, however, they are hardy, robust birds that cope well with both heat and cold. Although allowance must be made for their feathered legs—dry conditions are essential.

Confinement is never a problem, and the hens make good broody mothers, although their size can mean that broken eggs are sometimes an issue.

Gold Brahma male

Cochin

cuddly giant • profuse feathering • oriental roots • average layer

TYPE: large, heavy, soft feather • WEIGHT: male 10–13 lb (4.55–5.9 kg), female 9–11 lb (4.1–5 kg) • COLORS: Black, Blue, Buff, Cuckoo, Partridge, White

The Cochin is one of the most influential of all breeds, and many experts believe it to be the breed that spawned the whole poultry keeping and showing hobby that we have today. The Cochin—together with other exotic-looking, feather-legged breeds from Asia—was introduced to Britain in the mid-1800s, and it caused a sensation. People had never seen such large yet gentle chickens before. Until then, most fowl had been rather ordinary-looking creatures that were kept around the farmyard as little more than scavengers. However, the arrival of the enormous Cochin changed all that. Queen Victoria, who was a keen poultry fancier, is said to have been keeping "Cochin Chinas" as early as 1843, but it was not until 1850 that the breed really took off with the general public.

Unfortunately, today much of the original birds' egg-laying capability has been lost. Successive generations bred purely for exhibition purposes—when feather and fluff are maximized—mean that the breed's usefulness as a utility bird has been steadily eroded.

Looks

The Cochin body is large, deep, and short, giving the bird quite a square appearance. The tail is small and hardly shows at all in good examples because it is carried low.

The head is small, with large eyes and a small (especially on the females), serrated, single comb. The neck is short and covered with thick feathers, as are the legs.

The plain colors—Buff, White, and Black—are just that, with beaks that are, respectively, rich yellow, bright yellow, and yellow-horn-black. The Partridge version has attractive lacing, which is particularly apparent on the neck, wing, and back feathers. The Cuckoo variety features dark blue-gray bands across a lighter blue-gray background.

Personality

His good character is one of the Cochin's great strengths. Though not productive layers, the breed more than makes up for this with its friendly, docile nature.

This is a peaceful breed, which will live happily in most conditions and is very easy to handle.

Eggs

Despite the Cochin's size, the brown eggs are on the small side—big chickens do not necessarily lay big eggs. In general, you can expect about 110 a year from a healthy, contented young hen.

Day-to-day

Cochins are easy birds to look after as long as you have the right facilities. Their feathered legs mean that dry conditions are essential, and you will need to be careful about their feed because they are not active birds and will become fat if they are overfed.

They are not good foragers and will be perfectly happy if confined to a run, as long as they have adequate space. The hens make excellent broodies and are also content to act as stepmothers. They are slow-growing birds that are generally robust and hardy.

This is the "aristocrat" of the poultry breeds—large, majestic, and very content with life.

Black Cochin male

Croad Langshan

calm • good family bird • dark brown eggs • easy to keep

TYPE: large, heavy, soft feather • WEIGHT: large male 9 lb (4.1 kg), large female 7 lb (3.2 kg); bantam male 27–32 oz (770–910 g), bantam female 23–28 oz (650–790 g) • COLORS: Black, White

The Croad Langshan is another interesting Asiatic breed that first arrived in Britain in the early 1870s, when it was imported from the Langshan area in China by Major Croad, hence the name. Many experts thought that it was just a type of Black Cochin, but Major Croad persevered and, helped by his daughter, continued to breed the Langshan in its original form. This, appropriately enough, became known as the Croad Langshan. A taller version, known as the Modern Langshan, was also developed, but it failed to find lasting favor.

The breed never became as popular as the Cochin, particularly from an exhibition point of view, and this fact undoubtedly saved it from becoming a bird of fine feather but limited practical use.

The bright red single comb, wattles, and face give this breed striking yet simple looks.

Looks

Some people believe that the Croad Langshan bird is the most attractive Asiatic breed of all. He is a dignified, elegant bird, enhanced by his simple black or white coloring.

The males, in particular, appear to have very short backs because of the angle at which the tail is held. This creates an appealing U-shape between the neck and the tail.

Like other Asiatic breeds, the Langshan's head is small for the overall size of the bird. He has a single, upright comb with, ideally, five serrations. The face, earlobes, and wattles are bright red. The long legs are dark and feathered.

Personality

The Croad Langshan has a generally calm and well-balanced character, and the breed is a good choice for a family environment.

Eggs

Still retaining some of the breed's original utility abilities, the Croad Langshan is a reasonable layer of medium-sized, dark brown-colored eggs, which often have an attractive, plum-colored bloom on the shell.

Day-to-day

On a practical level, these birds are relatively easy to keep, although allowance must be made for their feathered legs—dry conditions are essential. In addition, they will flutter over low fences or hedges, so they must be kept within a secure boundary.

The breed is fairly hardy and, despite their Asiatic background, the birds are generally active and like to forage. They will tolerate confinement well, and the hens can make good broodies.

Black Croad Langshan bantam male

Dorking

ancient roots • docile • needs space • five toes • excellent table bird

TYPE: large, heavy, soft feather • WEIGHT: large male 10–14 lb (4.55–6.35 kg), large female 8–10 lb (3.6–4.55 kg); bantam male 2 1/2–3.0 lb (1.13–1.36 kg), bantam female 32–40 oz (910–1130 g) • COLORS: Cuckoo, Dark, Red, Silver-Grey, White

If you are thinking about buying a piece of British poultry history, you will not do much better than the Dorking. Records suggest that the breed has been around for 2,000 years or so, although their specific origin remains a mystery. Some people believe that the breed existed in Britain before the Romans arrived, but others suggest that the invaders brought it with them. Whatever the truth, the type we have today is far removed from the bird the Romans would have known. By the time the show-mad fanciers of Victorian Britain had finished fiddling with what, up until then, had been an extremely successful farmyard fowl, the Dorking breed was a mere shadow of its former self.

Looks

One of the best-known facts about the Dorking is that he has five toes on each foot, although no one really knows why. Apart from this, the breed is distinguished by a generally rectangular body that sits on short white legs. The males are large-bodied, impressive birds with large heads, combs, and wattles. The female birds are similar, although smaller overall than the males. They show long, horizontal backs, leading to a well-developed and broad-feathered tail (which is not fanned).

The colors available are Silver-Grey, Dark (single or rose comb), Red, White, and Cuckoo. Both the White and Cuckoo are rose-combed versions, and some experts consider the White to be the purest Dorking of all. There is also a Silver-Grey bantam version, but this is rare.

Red Dorking male

Dorkings love to forage and will require lots of space as they are very active.

Personality

The Dorking is a great breed for the domestic keeper, combining traditional good looks with an appealing temperament. The birds are calm and docile, easily tamed, and are generally always happy to be handled.

Eggs

You can expect 100–120 medium-sized eggs from a Dorking hen. These will be white-shelled from pure strains, but tinted if the birds have been crossed. In the past Dorkings were cross-bred with Sussex, and it was this that introduced the egg-tinting factor. Nevertheless, owners of Dark and Silver-Grey versions should find that their birds lay pure white eggs.

Day-to-day

The Dorking offers a fair source of eggs and tasty meat. The birds will enjoy foraging on a free-range basis, and they are not really happy in cramped conditions. If you have only limited space, perhaps the Dorking is not the breed for you.

They are reasonably hardy birds, and the hens make good broodies, although the chicks can be delicate and are slow to develop. For this reason, it is best to always hatch early in the spring so that the young birds can take full advantage of the warm summer months. This will give them the maximum amount of time to build up strength for the following winter. The breed has been widely used in the table poultry industry.

Faverolles

large and active • easy to live with • can be bullied • good layer

TYPE: large, heavy, soft feather • WEIGHT: large male 9–11 lb (4–5 kg), large female 7.5–9 1/2 lb (3.4–4.3 kg); bantam male 2 1/2–3.0 lb (1.13–1.36 kg), bantam female 32–40 oz (900–1130 g) • COLORS: Black, Buff, Cuckoo, Ermine, Laced Blue, Salmon, White

The attractive Faverolles is a good example of what is known as a composite breed—that is, one that was created by a program of careful crossbreeding. The influx of Asiatic fowl in the mid-1800s presented the most imaginative poultry breeders in Europe and North America with exciting opportunities. They crossed these exotic birds from the Far East with their existing stock to create a range of high-performing new breeds, including the Rhode Island Red, Wyandotte, Plymouth Rock, Barnevelder, Marans, and Welsummer.

Taking its name from the village in northern France where it was developed, this breed is thought to have resulted from the crossing of Cochin birds with Houdans and Dorkings. The breeders were trying to produce a heavy table bird that combined good eating quality with useful winter egg-laying capability. It has also been suggested that the Brahma and Malines were added into the mix as well.

The result—a robust, dual-purpose bird—first appeared in Britain in 1886, and as well as being appreciated in its own right, the breed was also used to cross with breeds such as the Sussex, Orpington, and Indian Game.

Looks

The Faverolles is a large, active bird with a broad, deep body. The head is broad, with a short beak, prominent eyes, a single, upright comb with serrations, face muffling (beard and whiskers), and small earlobes and wattles. The neck is always short, and the bird stands on shortish, feathered legs, with five toes on each foot.

There is a good range of plumage colors to choose from, with the Salmon variety being particularly attractive with its black, straw, and mahogany coloration. The Black has black eyes, beak, legs, and feet, and the feathers have a wonderfully rich beetle-green sheen in sunshine. The Buff is a lemon-buff color, with light eyes, beak, legs, and feet, and the Ermine is black and white.

The Laced Blue variety offers a combination of dark blue, laced with a darker shade, and the Cuckoo has a ground color of pale gray and darker banding on each feather. The White version is all-white.

Personality

The Faverolles usually has a great personality, and owners have said that these birds are calm, docile, and quiet, making them easy to handle and live with. However, the downside is that these birds can suffer from bullying, so if you intend to keep Faverolles in a mixed flock, keep watch for signs of trouble.

Eggs

The egg-laying performance of the Faverolles is reasonable. The light brown eggs aren't enormous, but keepers can expect to gather at least 100 from each bird in a year.

Day-to-day

As long as you avoid the potential bullying problem and you have good-quality, dry accommodations, you really should not experience many problems with Faverolles. They are hardy and early maturing, and they can make good broodies.

Salmon Faverolles females

Frizzle

striking looks • exhibition bird • good color range • labor intensive

TYPE: large, heavy, soft feather • WEIGHT: large male 8 lb (3.6 kg), large female 6 lb (2.7 kg); bantam male 24–28 oz (680–790 g), bantam female 20–24 oz (570–680 g) • COLORS: Black, Blue, Buff, Columbian, Duckwing, White, and many others

This breed was developed purely for exhibition purposes. The unique appearance comes from the feathers, which curl toward the bird's head in a most unusual way, making it the type of bird that you either love or hate. There is also some debate about whether the Frizzle should be considered a separate breed. The only country where it is classified as such is Britain; in most other countries it is simply regarded as a different feather formation, when "frizzled" versions of various breeds are recognized as little more than variations on a theme.

The origin of the distinctive feather type is not clear, although some records suggest that it was first encountered around 300 years ago in southern Asia. In Britain, both large and bantam versions of the Frizzle are available, and the bantams are much more popular than the larger birds. Even so, Frizzles in any form remain a rare sight. Large Frizzles have dwindled in numbers, although there are still a few specialist breeders working hard to preserve these impressive-looking birds.

Looks

Although they are not to everyone's taste, the curly feathers give this breed an unusual and interesting appearance.

The large males are strong, big birds that strut around looking imperious. They have broad but short bodies and big, loose tails. The head, which is dominated by a medium-sized comb and full, bright eyes, sits above an abundance of frizzled feathers on the neck. The legs are long and featherless, with four widely spaced toes on each foot. The comb is significantly smaller on the large female, which does not have so much frizzling on the neck. The bantam versions are similar to their larger counterparts.

There is a great range of colors to choose from, including some unusual forms, like the Cuckoo, Pyle, Spangle, and Red. In all cases, however, the comb, wattles, and earlobes are bright red. The

Highly frizzled feathers can look somewhat ragged, as on this bantam female.

legs and feet will tone in with the color of the beak, which varies between yellow, white, dark willow green, blue, or black, depending on the color of the plumage.

Personality

Frizzles are animated individuals that keep busy and active. The personality of frizzle-feathered versions of other breeds obviously matches that of the parent type.

Eggs

Frizzle hens will lay a reasonable number of white or tinted eggs.

Day-to-day

If you keep Frizzle bantams for fun, a group of mixed colors will live happily in the backyard. However, the males can be a little overbearing. The hens make fantastic broodies and mothers, but, for genetic reasons, the chicks produced tend to come in three feather types: flat, frizzle, and over-frizzled. The quality of feather is vital if you intend to show the birds, and careful, selective breeding is necessary (the flat-feathered version is a vital factor in the successful production of good exhibition stock).

Frizzles are hardy birds, perfectly suited to a free-range existence. They can also make excellent table birds.

Frizzle bantam female

Marans

chocolate brown eggs • active • gardener's friend • easy to keep

TYPE: large, heavy, soft feather • WEIGHT: large male 8 lb (3.6 kg), large female 7 lb (3.2 kg); bantam male 32 oz (910 g), bantam female 28 oz (790 g) • COLORS: Black, Dark Cuckoo, Golden Cuckoo, Silver Cuckoo

The Marans is a great breed to own and is probably best known for laying eggs with wonderful, chocolate brown shells. It was developed in France in the 1920s, and its original makeup is thought to include Malines, Croad Langshan, Faverolles, and Barred Rock, among others.

Originally bred in the region of western France around the town of Marans, not far from La Rochelle, the Marans is a dual-purpose breed, good for its meat and eggs. It first reached Britain in the late 1920s and, in its Cuckoo forms, has the big advantage of being sexable at an early age.

Looks

Marans are attractive chickens, with broad, deep and well-fleshed bodies. The head sits on top of a medium-length neck, and there are large, prominent, red-orange eyes (with big pupils) and a single comb with up to seven serrations. The light-colored beak is medium sized, and the red wattles and earlobes are similarly proportioned.

The bird stands on well-spaced, medium-length legs that are unfeathered and light in color. The feet have four toes and are clean and light in color.

The rare pure Black Marans has an appealing beetle-green sheen to his plumage, but the Cuckoo varieties are certainly more interesting and attractive to look at. They have an all-over pattern, with each feather banded with light and dark to create an extremely appealing overall effect. The Dark Cuckoo should show even, cuckoo banding (blue-black) across all parts, although a lighter colored neck is permitted. The Golden Cuckoo pattern shows a blue-gray coloring, with black and gold banding, and the Silver Cuckoo has white feathering on the neck and top of the breast and dark banding over a lighter ground color across the rest of the body. This gives a generally lighter impression than on the Dark Cuckoo version.

Personality

Marans prefer not to be handled and can be a bit variable when it comes to temperament, which seems to depend on the strain. It is worth checking this aspect carefully with the breeder before buying, especially if you have a young family. In general, however, they are busy birds that like to keep active. They're great to have around the garden and will happily pick away at slugs, snails, and other garden pests.

Eggs

These birds will certainly not disappoint as far as laying is concerned. The only potential problem is that, at a time when dark eggshells are so popular, some less than scrupulous breeders sell birds of dubious origins that, when they reach point-of-lay, do not actually produce dark eggs. Again, it is a case of buyer beware; check your suppliers carefully and buy only on recommendation.

Day-to-day

The Marans is a great forager and likes free-range conditions. It is a generally hardy breed and as such should present few problems if well kept. The best brown egg strains make fine broodies.

Cuckoo Marans male

Orpington

large and feathery • cuddly • can be bullied • reasonable layer

TYPE: large, heavy, soft feather • WEIGHT: large male 8+ lb (3.6+ kg), large female 6+ lb (2.7+ kg); bantam male 3$^{3}/_{4}$ lb (1.7 kg), bantam female 3$^{1}/_{2}$ lb (1.5 kg) • COLORS: Black, Blue, Buff, White

The Orpington is the classic example of a breed that has been dramatically changed, in both looks and performance, by its popularity as an exhibition bird.

Originally created in Britain in the late 1880s by William Cook and named after the town in Kent where he lived, it was the Black variety that emerged first. It is believed to have been produced by crossing Langshan, Minorca, and Plymouth Rock birds. Then, in the space of just eight years, Cook went on to create first the White, then the Buff forms. The Blue variety was not seen until the early 1920s.

Orpington bantams are simply scaled-down versions of the large fowl.

The dual-purpose birds that Cook would have known are a far cry from the feathery monsters we have today. Within just a few years of the creation of the Black variety, exhibitors started crossing it with Langshans and Cochins in an effort to boost what they regarded as its show potential. Unfortunately, although this made the Orpington bigger and more feathery, it also completely undermined its utility potential, and laying performance fell away dramatically.

Looks

The Orpington is one of those breeds that you either love or hate. The profusion of feathers is not to everyone's taste, particularly those who like to see more of a bird's body shape, but this bird is all about appearances, which is what sells him to an ever-increasing group of owners around the world.

They are bold and graceful birds with a deep, broad appearance. The wings are small, and the tail tends to appear short. The head is small, but with a strong beak and prominent eyes. The small, serrated comb is single, although the rose-type may be seen on the Black variety. The wattles are elongated but rounded, and the earlobes are small. The head sits on a shortish neck, which is made to appear even shorter by its full set of hackle feathers. The legs are short but strong, with the thighs usually virtually hidden from view by a mass of feathers.

The Buff is a popular color and should show even coloring all over, with red-orange eyes and a white beak and legs. The Blue is an attractive variation, with a variable slate blue ground color and delicate dark lacing on each feather. The legs and beak are dark, as are the eyes, although the toenails should be white. The Black and White versions are as their names suggest.

Buff Orpingtons, male (left) and female

Personality

The Orpington's personality matches his looks: he is cuddly and docile. These characteristics make him a great family bird and one that is tolerant of being handled, but his placid nature does means that he can be bullied by other birds.

Eggs

Even though the Orpington's egg-laying performance has suffered at the hands of those seeking success in the show pen, a healthy hen should still be perfectly capable of producing 160 good-sized, brown eggs in a season.

Day-to-day

The Orpington is a practical bird to own. These birds enjoy the chance to forage but will also tolerate confinement if necessary. They are pretty hardy, and the hens make good mothers. Do watch out for bullying if you decide to run these birds in a mixed flock.

Plymouth Rock

friendly • happy to be handled • good layer of large eggs

TYPE: large, heavy, soft feather • WEIGHT: large male 7½ lb (3.4 kg), large female 6½ lb (2.95 kg); bantam male 3 lb (1.36 kg), bantam female 2½ lb (1.13 kg) • COLORS: Barred, Black, Buff, Columbian, White

The Plymouth Rock is a popular, dual-purpose breed that was developed in Massachusetts, US, in the 1820s, but exactly how the first birds were produced remains something of a mystery. There are several theories, but the consensus is that Dominique males were crossed with either Black Cochin or Java females, although no one is absolutely certain.

The attractive Barred version of the breed was included in the first *American Book of Poultry Standards*, published in 1874. The White and Black varieties arrived next (as sports), then came the Buff version (created in Rhode Island). This was followed by the other color varieties, which were all developed in New England.

Looks

The Plymouth Rock is a big but well-balanced bird, with a deep body, straight back and medium-sized tail—everything is in proportion. The yellow beak is short and stout, and the large eyes are prominent. The single comb is serrated. The earlobes are large and tend to be about the same length as the wattles. The averagely long neck is slightly curved and covered with flowing hackle feathers, which extend down over the shoulders. There are no feathers on the legs, which are yellow, reasonably long, and well spaced. Each foot has four toes.

The Barred variety is an impressive-looking bird. The bright red comb and wattles contrast well with the attractive black-on-blue feather patterning. On good examples, every feather should have a black tip and even, clearly defined black barring (which includes the feather shafts).

The Buff is another popular color, and these birds should be evenly colored throughout the feathering, right down to the skin.

Personality

As far as personality is concerned, chickens do not get much better than the Plymouth Rock. The breed is known for its placid, docile, friendly temperament, and these birds are happy to be handled. They are friendly birds that are easily tamed and can make ideal pets for children.

Eggs

This breed is a good layer of large, tinted eggs, and healthy hens should produce 160 eggs in a season.

Day-to-day

The Plymouth Rock will be contented if confined, but equally happy to free-range if given the opportunity. He is not a particularly efficient eater and will mature reasonably early, so take care not to overfeed birds that are in smallish runs.

Above *White Plymouth Rock male*
Right *Buff Plymouth Rock bantam*

Rhode Island Red

ideal beginner's bird • fantastic layer • generally calm • robust

TYPE: large, heavy, soft feather • WEIGHT: large male 8 1/2 lb (3.85 kg), large female 6 1/2 lb (2.95 kg); bantam male 28–32 oz (790–910 g), bantam female 24–28 oz (680–790g) • COLORS: Rich Red-Black

This breed is another of those created in America by mixing indigenous fowl with imported birds from the Far East. It is perhaps the most successful and best known of the so-called composite breeds, and it is known and appreciated around the world for both its egg-laying and meat-producing qualities.

Work on developing the breed began just after the middle of the 19th century, when breeders in New England set about creating a truly dual-purpose bird by crossing what they already had with imported breeds, including Shanghais, Malays, Javas, and Leghorns. They made their selection on purely utility grounds, breeding the best males and females to maximize egg production and size. After following this simple, results-based program for many years, a pattern developed among the best-performing birds and a characteristic look emerged. The Rhode Island Red was born.

The first birds, which had rose combs, were exhibited at a poultry show in Massachusetts in 1880. A single-comb version soon followed (presumably as a result of introducing the Leghorn into the breeding mix), and by 1906 both types had been standardized in the US.

A specialist club for Rhode Island Red enthusiasts was established in Britain in 1909 and is still going strong. Joining really is a must for anyone new to the breed.

Looks

The Rhode Island Red is known for his generally rectangular profile. These birds are certainly more oblong than square, an impression that is enhanced by the long, flat back, only slightly raised tail, and vertical breast. However, the birds are not as red as they used to be or as the name suggests. Instead, modern birds are an attractive combination of rich reddish-brown and black.

The head is held slightly forward on a well-hackled neck and is topped with either a rose comb or a single, five-pointed, serrated comb. He has bright, prominent, red eyes. The earlobes and wattles are average size and finely textured.

The bright yellow legs are characteristic of the breed. They are well spaced and have four toes on each foot.

Personality

Rhode Island Reds have an almost ideal character for a productive backyard chicken. They are generally docile and calm birds (although the males can be a little feisty with each other) and will be happy to adapt themselves to whatever setup you can offer.

Rhode Island Red bantam female

Like any other breed, however, they will not tolerate overcrowding.

Eggs
Laying performance is the breed's strong point. A healthy hen should produce more than 200 large, brown eggs in a season.

Day-to-day
Rhode Island Reds are tolerant birds that will make the best of their surroundings, a trait that makes them particularly suitable for beginners. They are hardy and mature reasonably early. The only thing they do not do well is go broody, but, on just about every other count, the Rhode Island Red represents more or less the perfect bird for reliable egg production in the backyard. Sourcing good strains can be difficult, so take time to do your research if you are interested in this breed.

A pair of Rhode Island Red males in molt

Sussex

traditional favorite • brilliant layer • easy to handle • inquisitive

TYPE: large, heavy, soft feather • WEIGHT: large male 9 lb (4.1 kg), large female 7 lb (3.2 kg); bantam male 2 1/2 lb (1.13 kg), bantam female 28 oz (790 g) • COLORS: Brown, Buff, Light, Red, Silver, Speckled, White

If you are interested in combining the British poultry tradition with an excellent all-rounder, the Sussex should be your breed of choice. Few other breeds can offer the same combination of formidable laying performance, wonderful table qualities, and straightforward attractiveness.

The Sussex emerged in Britain in the 18th century, the product of crosses between indigenous, white-fleshed table fowl and newly imported Asiatic breeds, like the Brahma. It was present at the first poultry show in 1845.

Later, in its Light variety, the breed evolved into one of the world's best heavy-breed layers. In addition, the variety was capable, when crossed with "gold" males, of producing sex-linked chicks.

Today, the Sussex, with its seven standardized colors, is one of the more successful exhibition breeds, particularly in its popular bantam forms. The oldest of the color variants is the Speckled. The Red and Brown types arrived next (Old English Game was used in the Browns) and were followed by the Light (created using Brahma, Cochin, and Silver-Grey Dorking). The Buff appeared in the early 1920s, and the White arrived as a sport from the Light. The Silver is the most recent variety.

Looks

The Sussex is neat and well-proportioned. One of his trademark features is his wide, flat back, which, coupled with broad shoulders, a deep breast, and a tail held at an angle of 45 degrees, creates the classic look.

The head has a short, curved beak, good eyes, and a single comb, which is evenly serrated and vertical. The face, earlobes, and wattles are red and of a fine texture. The neck is pleasingly curved and has a good covering of hackle feathers, and the light-colored legs are featherless.

The most popular color variant is the Light, which has black-striped hackle feathers, a white body, and a black tail, a color scheme that is wonderfully finished by the bright red comb, face, wattles, and earlobes. The Buff has a similar overall look, but the white is replaced by a rich golden-buff.

The ground color of the Speckled is a rich mahogany, but each feather is also striped with black and tipped with white, giving a great overall effect. The tail is essentially black and white.

Personality

It is hard to fault the Sussex's character. The breed is docile, friendly, and easy to handle. At the same time, the birds are active and curious.

Eggs

A typical hen will lay about 180 good-sized, light brown eggs in a season, which is not quite up to the 220+ of the breed in its heyday, but is certainly good enough.

Day-to-day

The Sussex is a relatively easy and undemanding breed to keep, making them a sound choice for the beginner. The birds love to forage, but will be happy if contained. They are robust, tolerate low temperatures well, and make great broodies and attentive mothers.

Buff Sussex male

Wyandotte

fantastic layer • great temperament • attractive • hardy

TYPE: large, heavy, soft feather • WEIGHT: large male 9 lb (4.1 kg), large female 7 lb (3.2 kg); bantam male 3 3/4 lb (1.7 kg), bantam female 3 lb (1.36 kg) • COLORS: Barred, Black, Blue, Blue-Laced, Buff, Buff-Laced, Columbian, Gold-Laced, Partridge, Red, Silver-Laced, Silver-Penciled, White

For a comparatively young breed, the Wyandotte's dual-purpose history is surprisingly involved, which may be due to the wide range of color options. The Silver-Laced version was the first to be produced (in New York State) by crossing a Sebright cock with the offspring from a Silver-Spangled Hamburgh male—Cochin female combination. Further breeding resulted in a sample bird being put before the American Standards Committee in 1876, as an American Sebright, but this was rejected because the head and comb shape were considered by those on the panel to be wrong.

More crossbreeding followed, involving both Light and Dark Brahma females and a Silver-Penciled Hamburgh male. It also seems likely that Silver-Laced Polands were used as a source of lacing. In his *Illustrated Book of Poultry*, Lewis Wright reports that some of the early imports to Britain showed signs of a crest, which supports the idea that Polands were involved in the process.

The breeding program continued until 1883, when the resulting bird was finally standardized in America as the Wyandotte—named after a tribe of Native North Americans.

Once the Silver-Laced version arrived in Britain, breeders set about regularizing and improving the breed's lacings and ground color. Partridge Cochin and Gold-Spangled Hamburgh males were

Columbian Wyandotte bantams, male (foreground) and females

crossed with Silver-Laced females to produce the Gold-Laced variety. The Blue-Laced and Buff-Laced were created by crossing the Gold-Laced version with the White Wyandotte.

The White originated from sports of the Silver-Laced variety, and the broad and smooth-fitting feathering is an important factor of this type. The Columbian, which was named for the Columbian Exhibition held in Chicago in 1893, was created by crossing the White with a Barred Plymouth Rock.

Looks

The attractive Wyandotte is known for his curvaceous body, which is short and deep with a full breast. The saddle is wide and sweeps up to a fairly upright tail. The bird's head is broad, with a curved yellow or horn-colored beak, prominent red, orange, or bay colored eyes, a bright red rose comb, oblong earlobes, and average-sized wattles. The neck is of medium length and features plenty of hackle feathers. The legs are of medium length, with the thighs well-feathered but clean, preferably yellow, shanks (this yellow can become diluted on laying hens).

Of the many color versions available, the laced varieties are perhaps the most attractive, although the Partridge male is certainly a good-looking bird.

White Wyandotte male

Personality

The Wyandotte's character is well suited to the domestic environment. They are generally calm, friendly birds that are equally happy foraging outside or living in relative confinement.

Eggs

The Wyandotte is a great layer, and a good hen should produce 200 or more eggs in her first laying year, dropping to 175 in the second. Some colors lay better than others, with the Silver-Laced variety being the best layer.

Day-to-day

Looking after Wyandottes should be a straightforward business. It is a hardy breed, which is robust and will take temperature extremes in its stride. The hens make reasonable broodies and good mothers. Be careful not to overfeed Wyandottes, as they can become overweight.

Ancona

alert • boisterous • reasonable layer • experience recommended

TYPE: large, light, soft feather • WEIGHT: large male 6–6 1/2 lb (2.7–2.95 kg), large female 5–5 1/2 lb (2.25–2.5 kg); bantam male 20–24 oz (570–680 g), bantam female 18–22 oz (510–620 g) • COLOR: Black with white mottling

This striking-looking, clean-legged, large Mediterranean breed was named after the town of Ancona in Italy, where it was developed by crossing the local country fowl. The Ancona bears a striking resemblance to the Leghorn, and the debate about whether it should be classed as a member of that breed rumbles on.

The breed was first brought to Britain from Italy in the early 1850s, but it was another 30 years or so before it was taken across the Atlantic and introduced to breeders in America. It has proved popular in exhibition circles over the years, but has never really caught on widely with people who keep hens domestically. As a consequence, it remains a relatively rare breed outside enthusiast circles.

There are both large and bantam versions of the Ancona, the bantam being an exact, but smaller replica of its larger counterpart.

Looks

Technically, the Ancona is available in only one form—black with white, V-shaped mottling on the tips of the feathers—but some breeders in America have produced blue-white versions, although these have yet to be officially recognized as a standard color.

The Ancona can have two distinct comb types: single, with up to seven deep, broad serrations, or rose (similar to that found on a Wyandotte). The comb and large wattles should be bright red, and the bird will have pure white earlobes. There should be no feathering on the bird's legs, which are bright yellow with black mottling, and there are four toes on each foot.

Personality

Like other Mediterranean breeds, Anconas are flighty birds that are perfectly capable of fluttering over a low fence or hedge. This characteristic, combined with a generally excitable nature and a dislike of being handled too much, means that the Ancona is probably not best suited to life in a family environment, particularly if there are young children or other pets present. So all things considered, the Ancona is perhaps best suited to the experienced keeper.

Eggs

Anconas are good layers, and hens that are happy and healthy should reward you with 180 or so white-shelled eggs a year, which, in pure-breed terms, is pretty impressive. However, the eggs are usually on the small side.

Ancona bantam male

Day-to-day

Despite their boisterous nature, Anconas will tolerate confinement well, as long as the conditions are good. For best results, however, you must make sure that they have plenty of room and are kept amused with lots of distractions in the run to prevent boredom.

These birds love to forage given the chance—it suits their alert, active nature. They are also reasonably hardy, although they are not happy in very low temperatures. Anconas need to be effectively controlled, and if you are hoping to keep them in an open run, the fences will need to be high, which can add considerably to the setup costs.

Anconcas do not make good sitters, so if you are planning to try some natural brooding you will need a few other more suitable hens to be the broodies. If you are interested in rearing a cockerel or two for the table, bear in mind that although the breed matures quickly, the males can be a struggle to fatten up.

Ancona bantam female

Appenzeller

amazing V-shaped comb • white eggs • spectacular markings • very active • flyer

TYPE: light, soft feather • WEIGHT: Spitz male 3 1/2 –4 1/2 lb (1.6–2 kg), Spitz female 3–3 1/2 lb (1.3–1.6 kg) • COLORS: Black, Gold-Spangled, Silver-Spangled

The Appenzeller is a distinctive breed, with a split horn-type comb and forward-facing crest. It has been bred in Switzerland for hundreds of years and is named after the region in which it was developed. There are two versions—Spitzhauben and Barthuhner. The Barthuhner is the stockier of the two and has a rose comb (with backward-facing point) instead of the crest. It is also bearded and has white earlobes.

Looks

The Appenzeller has a spread, upright tail and dark brown eyes. The male Silver-Spangled Spitzhauben has silvery-white feathers, each with a small black "spangle" at the tip. The female is similar. The Gold-Spangled version has a gold-red ground color.

Personality

These are alert and nervous birds that are not terribly happy in close confinement. They will also fly given the opportunity.

Eggs

Reasonably good layers, they will produce 150 white eggs a year.

Day-to-day

The Appenzeller lays well and is fertile, but can suffer from a lack of robustness. The birds' flighty nature can make them hard to handle, and space is important. Their crests can cause problems in cold, wet weather, and care is needed to avoid frosting. They are not recommended for beginners.

Silver-Spangled Appenzeller female

Hamburgh

spectacular looks • flighty • not friendly • small eggs • not for beginners

TYPE: large, light, soft feather • WEIGHT: large male 5 lb (2.25 kg), large female 4 lb (1.8 kg); bantam male 24–28 oz (680–790 g), bantam female 22–26 oz (620–740 g) • COLORS: Black, Gold-Penciled, Gold-Spangled, Silver-Penciled, Silver-Spangled

Despite its name, the Hamburgh is believed to have been created in northern England over many hundreds of years. The precise origins of the breed are no longer certain, but it seems likely that a type of small, indigenous, pheasant-like fowl was crossed over a long period to produce what we have today.

Looks

The Hamburgh's long, sweeping tail complements his compact body, and the head is capped by a tapering rose comb with a fine, backward-facing spike, known as a leader.

The Black version has a distinct green sheen to his feathers, while the Gold-Penciled can be either a red-bay or a golden-chestnut color everywhere except for the tail, which is black. The sickle and covert feathers are laced with gold. The Silver version is the same, but with a silver ground color.

All types have red faces, combs, and wattles, white earlobes, and gray-blue legs and feet.

Silver-Spangled Hamburgh male

Personality

The Hamburgh is not an owner-friendly bird. In fact, they prefer to avoid human contact altogether if possible. This is a flighty breed, and the birds love to be left alone to forage. On the whole, they are best left to the experienced keeper.

Eggs

The Hamburgh is a good layer of small, white eggs.

Day-to-day

They are active birds and good flyers, and they are easily spooked. They are hardy, but the hens do not make good broodies, and the breed doesn't like close confinement.

Leghorn

large comb • prolific layer • white eggs • flighty

TYPE: medium, light, soft feather • WEIGHT: large male 7 1/2 lb (3.4 kg), large female 5 1/2 lb (2.5 kg); bantam male 36 oz (1.02 kg), bantam female 32 oz (910 g) • COLORS: Black, Brown, Buff, Cuckoo, Duckwing, Exchequer, Mottled, Partridge, Pyle, White

The Leghorn has proved to be an influential breed over the years. It originated in Italy, as one of the Mediterranean breeds with white earlobes—white lobes are always the sign of a good layer. It was distributed to several other European countries because of its laying prowess. But it was in America that the Leghorn was most extensively developed—a process that saw the breed become highly influential in many respects.

The first brown examples of the Leghorn were exported to the United States in the 1830s, and it is thought that they took their name from the Italian port of Livorno (Leghorn), from where the ships departed. White versions followed soon after, and it was these that were the first to come back across the Atlantic, this time to Britain, some 40 years later.

Breeders in the US used the Leghorn as the basis for a number of important breeding programs, including one that resulted in the creation of the Rhode Island Red. The breed has also played a crucial role in the development of many of the modern hybrid layers.

Looks

The Leghorn's appearance is dominated by his large single comb, which should always stand proudly upright on males but fall to one side on females. The birds have attractively proportioned, firm and well-feathered bodies, with reasonably full tails, long necks, and featherless legs.

Owners can choose from a good range of colors, with two of the most eye-catching being the multicolored yet traditional-looking Brown and the fine Exchequer. For those with simpler tastes, the Buff, White, and Black are more straightforward options.

No matter what their plumage color, all Leghorns have a yellow or horn-colored beak and red eyes. The face and wattles are bright red, the earlobes are pure white or occasionally cream-colored, and the legs are yellow or sometimes orange.

Personality

Like other Mediterranean breeds, Leghorns can be excitable and noisy. The bantams are more docile than the large versions.

Eggs

If you can live with their sprightly character, these birds will reward you with masses of eggs. Leghorns are one of the top pure-breed layers, and each healthy, contented hen should produce about 200 eggs a year.

Day-to-day

Leghorns can cope well with confinement and are good winter layers. They can be flyers, so you will need to think carefully about providing a secure enclosure. It is probably best, too, if it is sited away from the house to keep to a minimum any potential causes of disturbance to the birds.

The hens are nonsitters, so do not even bother trying any natural brooding. They do love to forage and are generally robust, often living for over six years. Be warned, however, that their large combs can suffer damage in very cold weather.

Brown Leghorn male

Minorca

restless nature • good layer • large comb • graceful looks

TYPE: large, light, soft feather • WEIGHT: large male 7–8 lb (3.2–3.6 kg), large female 6–8 lb (2.7–3.6 kg); bantam male 34 oz (960 g), bantam female 30 oz (850 g) • COLORS: Black, Blue, White

The Minorca was an established Mediterranean breed for well over a century before it began to achieve any degree of popularity in Britain. It had been favored for years in the west of England, but it wasn't until the white-faced Spanish started to fall from grace in the 1870s that the red-faced Minorca really came into its own as a good egg-layer.

The breed's growth in popularity was something of a double-edged sword, however, because it became popular as an exhibition fowl and started being selectively bred to develop its head features. This, somewhat inevitably, had a detrimental effect on the breed's egg-laying performance.

Looks

The Minorca is a graceful-looking bird. His upright stance, angled body, and full tail give him an attractive profile, and the Black version is especially sleek.

The head is dominated by the comb. In its single form on the male, it is large and deeply serrated (ideally with five points). The female's comb, instead of being upright like the male's, falls to one side. It does not matter which way it goes, but the important thing is that it should not obstruct the hen's vision by hanging down too low.

There is also a version with a rose comb, which is covered in small nodules and features a backward-facing leader. The eyes are full and bright, while the other most striking

Pair of elegant Black Minorcas

features are the large, elongated, white earlobes, which can be nearly 3 inches (8 cm) long on male birds, and the large, rounded red wattles.

The popular Black Minorca has glossy black feathers, a dark beak, eyes, and legs, and four toes on each foot. The White and Blue types, which are less common, have, respectively, glossy white and soft blue plumage, with light and blue legs.

Personality

The breed's Mediterranean roots mean that the Minorca can be a handful for inexperienced keepers.

Eggs

A Minorca hen in good condition will produce about 200 large, white-shelled eggs in a year.

Day-to-day

The Minorca tolerates hot weather well, but the size of the single combs can pose problems in cold, frosty weather because they may suffer from frostbite.

The breed's restless nature means that secure fencing will be necessary. Although the birds mature quite quickly, the hens do not make good sitters.

If you can live with the Mediterranean temperament, the Minorca's graceful appearance will enhance any setting.

Black Minorca male

Poland

striking looks • high maintenance • exhibition favorite • friendly

TYPE: large, light, soft feather • WEIGHT: large male 6 1/2 lb (2.95 kg), large female 5 lb (2.25 kg); bantam male 24–28 oz (680–790 g), bantam female 18–24 oz (510–680 g) • COLORS: Black Crested White, Blue, Chamois, Cuckoo and Buff, Gold, Self Black, Self White, Silver, White Crested Black and White, and Cuckoo

Silver-Laced frizzle Poland male

Many people believe that the Poland is one of the oldest breeds in existence, and there is some evidence to suggest that it predates the Romans, although no one is really sure. As with so many of the ancient poultry breeds, confusion and contradictory theories surround the Poland's origins.

One of the few things that is fairly certain is that the breed does not actually have any direct link with Poland. Its roots lie elsewhere in Europe, and there are historical references to it in the Netherlands and France. As far as the name is concerned, one theory has it that it arose simply as the result of a series of corruptions of the original name, Poll. Over the decades this word, which refers to the breed's characteristic and impressive head crest, evolved through Pole and Polled before finally becoming Poland.

Although these birds can be reasonably good summer layers, they are rarely kept for this reason today. Most owners love their Polands simply for the way they look, and they are a popular exhibition breed. The breed was represented at Britain's first poultry show, held in London in 1845, and appeared in the first edition of the book of *British Poultry Standards*, which was published in 1865.

Looks

The Poland's appearance is dominated by his crest, the feathers of which grow out of a protuberance on top of the head. The color of the crest can either match the rest of the bird's plumage or be a contrasting white.

The bird himself is bold and active, with a long back and full breast. The tail should be full and neatly spread but never upright. The large head is mostly hidden by the crest, which should be evenly balanced when viewed from the front.

The comb, which is a V-shaped (horn) type, should be very small, but the nostrils on either side of the beak are large. White-Crested Black birds should have no face muffling (beard and whiskers) and long wattles, while the rest all have abundant muffling, which hides the small earlobes, and no wattles at all.

There are more than ten plumage colors to choose from, but the most commonly shown type is the White-Crested Black.

Personality

Today, most Polands are kept for exhibition purposes or by fanciers who appreciate their appearance and beauty. They have a friendly nature, are easy to handle, and make good pets.

A pair of White-Crested Black Polands

Eggs

You will get about 120 smallish, white-shelled eggs from a Poland hen in a season.

Day-to-day

The Poland's fancy head feathering can cause problems at a practical level if he is not looked after properly, making it not an ideal starter breed. The crest must be inspected and washed regularly to prevent parasitic infestations, which can cause serious problems, including blindness. In addition, because the crest limits the bird's vision, the breed is not suited to living in a mixed flock as other, noncrested breeds can take advantage of him.

The birds should always use narrow-lipped drinkers to prevent their head feathers from getting wet, and they should be fed on pellets rather than mash because dust from mash can cause eye problems.

Scots Dumpy

Scottish roots • docile • good table bird • tricky to breed

TYPE: large, light, soft feather • WEIGHT: large male 7 lb (3.2 kg), large female 6 lb (2.7 kg); bantam male 28 oz (790 g), bantam female 24 oz (680 g) • COLORS: various, including Black, Brown, Cuckoo, Gold, Silver, White

The Scots Dumpy is aptly named: the breed originated in Scotland and is known for its extremely short legs.

Records show that the breed existed in Scotland as early as the 1700s and that one of the first places it appeared in England was Newmarket in 1852. The breed came close to extinction in the late 1800s, when it was reported that they were being kept by only a few people in Scotland. Now, however, the formation of a club dedicated to the breed's survival means that the Scots Dumpy's future is assured.

The odd combination of a long, deep body and short legs has given rise to all sorts of interesting names over the years, including Creepers, Crawlers, and Bakies. Similarly, short-legged breeds can also be found in France and Germany.

Looks

The breed's most striking feature is, as the name suggests, its short legs. The fact that the adult bird is

Scots Dumpy male

A Scots Dumpy pair; these are very docile birds, but it is tricky to breed good examples.

usually less than 2 inches (5 cm) off the ground causes him to adopt an unusual, waddling gait.

Despite the short legs, the body is large, long, and broad. The back is flat, and the tail is full, with well-arched sickle feathers on male birds. The head features a strong, curved beak, large red eyes, and a single, serrated comb. The face is smooth and red, as are the averagely sized wattles and earlobes. The neck is covered in long, flowing hackle feathers.

There aren't actually any clearly defined colors for the Scots Dumpy, although those listed above are the most often found.

Personality

These birds, which are very likeable, are reasonably docile and can make good backyard fowl for keepers with a bit of experience.

Eggs

The Scots Dumpy is a good layer of light-colored eggs.

Day-to-day

The breed's short legs can be a cause for concern from a breeding point of view. There is a tendency for the breed to revert back to a longer legged form, and if you try to prevent this by crossing short-legged versions with each other you run the risk of increasing infertility and mortality rates. However, the Scots Dumpy hen is known for her excellent mothering skills, and the breed is also said to make a very tasty table bird.

Scots Grey

proud • excitable • elegant • robust • fair layer

TYPE: large, light, soft feather • WEIGHT: large male 7 lb (3.2 kg), large female 5 lb (2.25 kg); bantam male 22–24 oz (620–680 g), bantam female 18–20 oz (510–570 g) • COLOR: Barred Black-Grey

The Scots Grey used to be known as the Scotch Grey and before that, colloquially, as the Chick Marley. It is an interesting and elegant bird that has plenty to offer keepers with a bit of space and a liking for a rare but good all-rounder.

The breed has been around for at least a couple of centuries in its native Scotland, and it seems to have been developed essentially from farmyard fowl. Records reveal that it was a popular show bird in the early and mid-1800s, although typical examples were smaller and more coarsely marked then than they are today.

The breed fell from favor toward the end of the 19th century, and a club for keepers and enthusiasts was formed in 1885 in a bid to help preserve it. The Scots Grey is thought to have had some game blood in its original makeup, which contributed to its length of leg, but no one is certain about this.

Nowadays, although it is still something of a rarity, the breed's future is assured, which is good news for people who are looking for something a bit different.

Looks

Both males and females are striking-looking birds. They are active, upright, and bold, with broad, flat backs, deep, prominent breasts, and fairly long wings. The strong beak is predominantly light in color. The large eyes are amber, and the face, comb, rounded wattles, and earlobes are bright red. The medium-sized comb is a single type, which is held upright and shows up to six sharp serrations.

The neck is attractively tapered, and long hackle feathers extend down over the shoulders on the male. The bird stands on strong, long, widely spaced legs, which are light-colored and featherless. Each foot has four toes.

The bird's attractive plumage is the result of a steel grey ground and distinct black barring across all feathers. The evenness of these markings combined with the absence of white or any rustiness in the plumage are vital features from an exhibitor's point of view.

Personality

These are proud birds with lots of character, but their active nature means that they can sometimes be rather excitable when disturbed. Males tend to have a feisty attitude during the breeding season. Therefore, this is not a breed best suited to life in a family environment, particularly if there are young children present.

Eggs

The Scots Grey was developed as a country-type utility fowl, so you can expect fair numbers of good-sized, light-colored eggs from hens that are well cared for.

Day-to-day

Its Scottish roots have made sure that this breed is hardy and a good forager. However, these birds are not best suited to confinement and can react badly by feather pecking and the like when kept where space is limited. The hens are reliable sitters. A Scots Grey bird matures reasonably quickly and is said to make a tasty roast.

Scots Grey male

Silkie

unique looks • gentle and calm • wonderful broody • needs TLC

TYPE: large, light, soft feather • WEIGHT: large male 4 lb (1.8 kg), large female 3 lb (1.36 kg); bantam male 22 oz (620 g), bantam female 18 oz (500 g) • COLORS: Black, Blue, Gold, Partridge, White

The Silkie is another breed that really is unlike any other chicken that you are likely to come across. Their most obvious feature is the strange, fur-like feathering, which gives them an unusual, some would say almost comical, appearance. They resemble a ball of fluff, with feathered legs, and, for good measure, a powder puff-like crest on top of the head. If that were not enough, there are also bearded versions, which have ear muffs as well as beards.

These are not the only unusual features of this ancient Asian breed. The Silkie is one of an exclusive group of breeds that features five toes on each foot. It also has dark purple (almost black) skin and similarly colored comb and wattles, and the earlobes are blue. The dark skin means that these birds look rather odd when they are cooked, an impression that is heightened when it is found that the bones, too, are almost black.

Like many of the other old poultry breeds, nobody is quite sure about the origin of the Silkie, but it is widely accepted that the bird has been around for many hundreds of years—a "furry" chicken was documented by Marco Polo during his travels in China in the late 13th century. Trade routes set up between the East and West brought the Silkie to Europe, and records show that in the Netherlands the birds were sold as the product of crossing a chicken with a rabbit.

Even though the Silkie is officially classified as a large fowl, its

White Silkie male

diminutive size means that people often mistake the birds for bantams. The genuine bantam version, which was finally standardized in Britain in 1993, is, in fact, about two-thirds as small again.

Looks

The male Silkie has a broad, stout body beneath his mass of fluffy feathers. The short back leads to a ragged-looking tail, and the wings also have a ragged appearance, particularly the tips of the flight feathers, which hang down in a series of straggles.

The head is dominated by the fluffy crest feathers, but also features a dark, walnut-type comb, black eyes, a short beak, and mulberry-colored face and wattles. The earlobes should ideally be a shade of turquoise blue, but a similar color to the rest of the head gear is permitted.

The legs, which should be dark gray, are moderately feathered, as are the middle and outer toes on each foot. The female Silkie shows more cushioning on the saddle, so that the tail is almost covered. The legs are shorter, too, so the bird's under-fluff almost touches the ground, and the comb, earlobes, and wattles are all smaller.

Silkies have so much going for them: unique looks, good character, and tremendous broodiness.

Personality

The Silkie has a wonderfully gentle and calm character, making these birds ideally suited to being kept in the backyard. However, they do not mix well with other birds and are best kept on their own.

Eggs

The fact that Silkie hens spend so much time being broody eats into their egg-laying potential. You will be lucky to get close to 100 smallish eggs a year from a good bird.

Day-to-day

Two factors make the Silkie a particularly good bird for the hobby keeper: they don't fly, and they are quite happy to live in reasonably close confinement. A disadvantage of the breed, however, is that the unusual feathering makes it susceptible to wet or muddy conditions, and scaly leg can be made worse by the feathered legs. Feeders and drinkers must be carefully chosen to prevent crests from getting wet or dirty, which, in turn, can lead to parasitic infestation and sometimes eye problems. Marek's disease can be an issue, so check carefully when sourcing stock.

Welsummer

good layer • dark eggs • child-friendly • docile • poor mother

TYPE: large, light, soft feather • WEIGHT: large male 7 lb (3.2 kg), large female 6 lb (2.7 kg); bantam male 36 oz (1.02 kg), bantam female 28 oz (790 g) • COLORS: Red-Brown, Silver Duckwing

Named after the village of Welsum in the east of Holland, where it was developed, the Welsummer was added to the British Poultry Standards in 1930. The breed's large, "flowerpot brown" eggs were an instant hit. The eggs were so brown that some people thought they had been faked, an idea that was not helped by the fact that the pigmentation has a tendency to wash off newly laid eggs.

It is reported that Partridge versions of the Cochin, Wyandotte, and Leghorn were used in the early breeding process and that all sorts of oddities gave rise, which included birds with bright yellow feathers, five-toed feet, and blue tail plumage.

Nevertheless, the bird showed great promise, and the breeders persisted. It is likely that both the Barnevelder and Rhode Island Red played important parts in this, helping to give the Welsummer its impressive egg-laying potential.

Looks

The Welsummer is an attractive if not spectacular bird, and the males have the traditional farmyard cockerel look about them. Both sexes are active and adopt a generally upright, busy manner. They have a long back, but have a characteristic U-shape, formed by the neck, back, and tail.

The well-proportioned head features a short, light-colored beak, alert, orange eyes, a single, upright comb with between five and seven serrations, a smooth red face, small red earlobes, and averagely sized red wattles. The legs are featherless and yellow, and the feet have four well-spread toes. Male birds show more color than the females and are an attractive mixture of reds, brown, and black with a beetle-green sheen. The hens are much plainer, being essentially brown with black speckling (a partridge-type marking), although the hackle feathers are gold-shafted.

Welsummer large fowl and bantam females together

A Duckwing version is mentioned in the breed standard, and this is essentially a black and white bird, although the hens do have a salmon-red breast and silver-gray coloring elsewhere.

Personality
The Welsummer has a friendly and docile character, and they are easy to handle and child-friendly, making them ideal as backyard birds. Their calm temperament makes it a pleasure to own in all respects.

Eggs
A good, healthy hen will produce between 140–160 eggs in a season, all with that wonderfully rich brown shell.

Day-to-day
These birds do almost everything right. They are easy to look after and will be happy foraging or living in confinement. They are hardy and economical eaters. Just about the only negative aspect is that the hens do not make good mothers, and if you want to hatch eggs you will need an incubator.

Welsummer pullets

Belgian Bantam

tiny • active • decorative • can be aggressive

TYPE: true bantam • WEIGHT: male 24–28 oz (680–790 g), female 20–24 oz (570–680 g) • COLORS: Many and varied

Originating in Belgium in the early 1900s, these are true bantams, which means that they have no large fowl equivalent. There are three main types: the Barbu d'Uccle, the Barbu d'Anvers, and the Barbu de Watermael.

Looks

The Barbu d'Uccle has a great neck hackle and a well-developed breast. The wings point down at an angle. The short, curved beak sits above the famous beard.

The Barbu d'Anvers also has a large neck hackle. The head appears big, and the beak is two-tone.

The Barbu de Watermael differs because of his crest and rose comb, which combine to make the head look larger than it really is.

Personality

They have friendly and engaging characters, although some of the cockerels can be a little aggressive during the breeding season.

Eggs

You won't keep Belgian bantams if you want eggs. Those that are produced are tiny and a creamy-white color.

Day-to-day

Although these birds are tiny, they are very active and good fliers, so they will need an effectively fenced run. They will put up with confinement, but also like to forage.

Although they are relatively hardy, Belgian bantams need to be protected from bad weather.

Quail Belgian Barbu d'Anvers male

Booted Bantam

proud • feathered legs • docile • good free-ranger

TYPE: true bantam • WEIGHT: male 30 oz (850 g), female 27 oz (750 g) • COLORS: Barred, Black, Buff, Columbian, Millefleur, Partridge, Pearl Grey, Porcelaine, White

One of a handful of ancient, true bantams, the Dutch Booted Bantam is popular in mainland Europe. It is known to have been kept in the Netherlands since the 17th century. The Booted part of the name refers to the long vulture hocks (formed by long, stiff quill feathers at the hock joint) that characterize the breed.

Looks

Booted Bantams are proud, thick-set birds. The breast is high and carried forward, while the neck and head are angled slightly backward. The body is short and compact, with a broad back and wings that are held low. The saddle feathers are long and abundant.

On males there is a well-serrated, five-pointed comb, which is red, as are the earlobes and wattles.

Personality

These birds are generally friendly and docile. They are perfectly happy in smaller houses, although, of course, the bigger the house the better. They are also great to run in the garden, because their foot feathering means that they do not cause too much damage to plants.

Eggs

Egg production is good in summer, with the color of the tiny eggs varying between white and tinted.

Day-to-day

The foot feathering requires constant attention. The henhouse must be kept clean, otherwise these feathers quickly become dirty or, worse still, stick together. The foot feathering and the fact that they are nonflyers mean that these birds require wider and lower roosting perches than usual. Generally, though, the breed is hardy.

These birds are known as sitters, but not all varieties can be relied on, and inexperienced keepers may find chicks hard to rear.

Millefleur Booted Bantam female

Dutch Bantam

simple to care for • friendly • colorful • easy to handle • hardy

TYPE: true bantam • WEIGHT: male 18–20 oz (510–570 g), female 14–16 oz (400–450 g) • COLORS: Black, Buff, Buff Columbian, Cuckoo, Lavender, Millefleur, Pyle, Salmon, Silver Duckwing, Wheaten, White, and many others

The Dutch is another of the handful of true bantams—genuinely small birds with no large fowl counterpart; it is a breed in its own right. One of the smallest and most friendly bantams you will find, it is available in more than 20 different varieties.

Not only is this breed very attractive, it is practical too, being capable of laying more than 160 eggs in a year and being able to thrive even when space is limited. Consequently, the bird is well suited to the domestic environment.

The Dutch Bantam's background is not entirely clear. Partridge-colored "farmers' bantams" have been a common sight in farmyards across the Netherlands for many hundreds of years, and it is likely that these were the ancestors of today's Dutch. The first known documented reference to the breed came in 1882, when it was mentioned in a handbook by the managing director of a zoo in The Hague. Originally, it was available in just one variety—the Partridge, which is sometimes referred to as the wild color or Bankiva. The breed has historical links with other old Dutch breeds, including the Friesian, and was first officially recognized by the Dutch Poultry Club in 1906.

Dutch Bantams found their way to Britain in the early 1970s, a club was formed in 1982 (13 colors are standardized), and the breed has grown in popularity all around the world.

Gold Dutch Bantam female

Looks

The Dutch Bantam is an upright, active little bird with a short, almost U-shaped back and plenty

Pair of Gold Dutch Bantams

of feathering. The ornamental feathers (tail sickles, coverts, side hangers, and hackles) should all be well developed.

The head, which sits on a short, curved neck, is relatively small with a short, horn-colored beak and a red, single comb with five small serrations. There are small, red and rounded wattles, and these contrast with the white, almond-shaped earlobes. Eyes can range in color from orange to dark brown. The relatively large, well-rounded wings are carried low and backwards. The upright and large tail should be full, well spread, and carried high. The cockerels have well-developed, curved sickle feathers.

The bird stands on rather short and well-spread legs, which vary in color from dark to light blue slate.

Personality

This is a good bird to have around. They are easy to handle and can be trained to be very friendly.

Eggs

Owners are often surprised by what good layers and good sitters Dutch Bantams are. You can expect each bird to produce around 165 light-colored eggs a year, although they tend to be small.

Day-to-day

There are few problems to worry about with these birds. They tolerate confinement well, but they can fly, so you will need to think carefully about suitable enclosures. They are generally hardy and will forage happily if allowed to do so. The male birds can be a bit feisty.

If space is limited, the Dutch Bantam is certainly a breed that will prove easy to live with.

Japanese Bantam

decorative • tiny • dove-like • some fly • good mother

TYPE: **true bantam** • WEIGHT: **male 18–20 oz (510–570 g), female 14–18 oz (400–510 g)** • COLORS: **Birchen Grey, Black, Black-Tailed White, Blue, Cuckoo, Golden Duckwing, Mottled, Silver Duckwing, Silver-Grey, White and others**

The Japanese Bantam is another of the tiny breeds classified as a true bantam, which means that it is a breed in its own right and that there is no large equivalent. Most bantams are simply hybridized, miniature versions of a full-sized breed, but, as with other true bantams, such as the Booted Bantam, Rosecomb, and Sebright, the Japanese Bantam is a distinct and fascinating breed.

Apart from the wide range of plumage color possibilities available, another thing that sets the Japanese Bantam apart is that it has the shortest legs of any standardized breed. The legs are

Black-Tailed White Japanese Bantam male

so short, in fact, that you often cannot see them, and the birds appear to walk with a definite waddling movement.

There are three basic feather types, which make a big difference to the bird's overall appearance. The plain-feathered birds have straightforward, normal-looking feathering. The slightly more unusual "silkie" type have soft, flowing plumage. But perhaps most eye-catching of all are the frizzle versions, which have strangely curled feathers, the tips of which tend to bend back on themselves and point toward the bird's head.

The breed is thought to have been initially developed in Japan as early as the 7th century, but it was not seen in Europe until the 1860s; since then its type has remained unchanged.

Frizzled Black-Tailed White Japanese Bantam female

Looks

These are tiny birds, and the males in particular can seem a little top-heavy with their upright, full tails and large, bright red, single combs. They all have short backs, and full and proud breasts. Looked at from the side, the best examples can show almost no gap between the base of the neck and the base of the tail—ideally there should be a tiny U-shape here.

The wings, which are long in relation to the rest of the body, are angled so that the tips nearly touch the ground close to the back of the body. The spectacular tails rise up well above the height of the head and comb. There are impressive, sword-like main sickle feathers plus a number of softer side hangers.

The head is dominated by large eyes (often red, sometimes orange) and by the impressive comb, which should be evenly serrated with no more than five points. The short legs are sharply angled at the joint and are not feathered. There are four, well-spaced toes on each foot.

Personality

If you like small birds, you will probably love this breed. They are generally friendly little birds, although males can be aggressive in the breeding season.

Eggs

The breed is not known for its laying ability. The eggs that are produced have cream shells and are tiny.

Day-to-day

These birds require care and attention. Do not be tempted to overcrowd them because they are so small—no breed likes that—and their fancy feathering and low stance mean that they must be kept in clean, dry conditions if they are to prosper.

They are active birds, and some can be fliers, so they will need to be enclosed carefully. The hens make excellent and attentive mothers and superb broodies. Fertility can be a problem due to shortness of the leg.

Pekin

great personality • ideal family bird • plenty of colors

TYPE: true bantam • WEIGHT: male 24 oz (680 g), female 20 oz (570 g) • COLORS: Barred, Black, Blue, Buff, Columbian, Cuckoo, Lavender, Mottled, Partridge, White

Pekin bantams are known in the rest of the world as Miniature Cochins or Cochin Bantams, although this is confusing because it is generally agreed that the Pekin has no connection with the Cochin—apart from their similar origins—and that it is a breed in its own right. The Pekin is not an exact miniature version of the Cochin.

The first Buff Pekins arrived in Britain after having been "liberated" from Peking (Beijing) by the British Army in 1860. They came at a time when many of the Cochin-type birds from elsewhere in China were being described as Shanghais, and this led to a certain amount of confusion. In time, Cochin became the accepted generic term for all birds of this type, and it is probably this history that lies at the root of the supposed Cochin connection. The situation became further complicated by the importation of the much more profusely feathered and larger Cochin bantams from America and then, more recently, a steady stream of similar birds from mainland Europe.

As with so many other breeds, today's Pekin is thought to bear little resemblance to the type that was brought from China nearly 150 years ago. Those original birds were taller, more upright, harder feathered, and generally bigger. In fact, it would be hard to make them more different if you tried.

Columbian (left) and Frizzled Lavender (right) Pekin females

Today's Pekin performs a dual role. It is an extremely popular garden pet, but at the other extreme it is an exhibition bird that requires considerable breeding skill and knowledge.

Looks

The Pekin has been likened to a ball of fluff on legs, which is not far from the truth. A good example will have a short, broad body, and the head will be set only slightly lower than the top of the tail. This gives the bird a characteristic tilt forward, which is an important feature in an exhibition bird.

The short legs are feathered and barely visible. The small head is dominated by a single, well-serrated comb. The bird's face, earlobes, and wattles are all smooth, and on good examples the lobes and wattles are nearly as long as each other. The head sits on a short neck, and there is an abundance of soft hackle feathers, reaching well on to the back. There are curves everywhere on the Pekin, and there should be no stiff feathers in sight, although those on the legs are stiffer than those on the body.

That there are so many colors available and that more seem to be appearing on a regular basis are matters of great concern to serious enthusiasts of the breed. They feel that the standard colors are being put at ever-increasing risk by unscrupulous breeders who are simply seeking to create new color combinations to satisfy demand from the pet keepers.

Black, Lavender, and Buff Pekin females

Personality

Pekins are great birds to have around the garden. They have docile yet cheeky characters, and the youngsters can be real fun.

Eggs

Considering their size, the Pekin is a pretty reasonable layer of good-sized, tinted eggs.

Day-to-day

Pekins are great birds to keep if you are looking for something that is decorative, friendly, and full of personality. Their size makes them ideally suited to environments where space is limited, although males can sometimes get a bit peppery. Their feathered feet mean that they must be kept in clean, dry conditions if they are to prosper.

The hens make great mothers and regularly go broody, thus making Pekins an extremely useful breed for natural incubation.

Rosecomb

exhibition beauty • poor layer • friendly • difficult to breed

TYPE: true bantam • WEIGHT: male 20–22 oz (570–620 g), female 16–18 oz (450–510 g) • COLORS: Black, Blue, White

Being a true bantam, the Rosecomb has no large breed equivalent. Traditionally, these birds have been kept and perfected by people who are keen on showing. Details about the breed's origins are sketchy, and there are several theories about where the breed came from. It has been suggested that it was brought to Europe from the Far East (Java), but other people claim that it is little more than a miniature Hamburgh, developed in the Netherlands. Another theory holds that it is a genuine British breed that has existed and been carefully bred in Britain for hundreds of years, and there is some evidence to support this last view, with literary references suggesting that the breed was around in the late 1400s. Ultimately, however, nobody can be completely sure.

Whatever its origins, the breed's survival seems to be entirely due to its beauty, because it has no real utility value at all.

Looks

The Rosecomb is officially described as a "cobby" bird, which is an old-fashioned word meaning short and compact. It is a jaunty little creature, with a prominent, well-rounded breast, a short back and wings that are angled toward the ground (more so on the male) and cover much of the thigh.

However, it is the bird's head that is of the greatest importance from an exhibition point of view. The beak is short, and the rose-type comb features a long leader, which extends out toward the bird's tail. The round earlobes, which should be pure white, are much larger on males than on females. Wattles are fine and rounded and, once again, much smaller on female birds. The neck is short and attractively curved, with an abundance of hackle feathers, which can be long enough to reach the base of the tail. The short legs are dark gray and featherless.

Personality

Rosecombs are generally friendly little birds, although the males can be a little aggressive during the breeding season if provoked. Overall, though, they are engaging characters with plenty of personality.

Eggs

Laying is not one of the breed's strong characteristics, although they lay quite well in the summer.

Day-to-day

These birds are generally happy to tolerate confinement when required, but they can fly, so if you intend to let them free-range around the garden—they love to forage—you should think carefully about the height of the fences needed to keep them in.

Breeding is a difficult business for the inexperienced keeper. Infertility can be a problem, and some say that this is a consequence of all the inbreeding that has taken place over the years to keep the breed pure and up to standard.

Black Rosecomb male

Sebright

fantastic looks • flyer • delicate chicks • exhibitor's favorite

TYPE: true bantam • WEIGHT: male 22 oz (620 g), female 18 oz (510 g) • COLORS: Gold, Silver

Gold Sebright male

This absolutely stunning true bantam is a real favorite with exhibitors and hobby keepers alike simply because of its wonderful laced markings. It is impossible not to love the crisp markings of this jaunty little bird.

The Sebright was developed and named by Sir John Sebright just over 200 years ago. He set out with the straightforward aim of producing a laced bantam, and he also established a club for his creation in about 1810. The actual breeds he used in the development process remain something of a mystery, but popular belief has it that Poland, Nankin, and Hamburgh stock was involved and that Sebright used the Rosecomb bantam of the period as his starting point for breeding.

An interesting feature of the breed is that male birds are what is known as hen-feathered—that is, they don't develop the large, curved sickle feathers in their tales that are so typical of most other male chickens. This feature is thought to have a detrimental effect on fertility. Consequently, some breeders recommend that those males that do happen to show any signs of growing sickle feathers are used in the breeding pen.

Over the years, the color of the Sebright's comb, wattles, and earlobes has changed from a deep purple color—termed gypsy-faced—to a more conventional, redder shade (described in the breed standard as mulberry).

Silver Sebrights enjoying themselves in deep litter.

Looks

Despite their diminutive size, Sebrights strut self-importantly. They are compact birds, with short backs and prominent breasts. The wings, which are large for the overall size of the bird, are carried low and angled toward the ground. The tail should be well spread and carried at a high angle. As well as causing a lack of sickle feathers, the male Sebright's hen-feathering also means that there is an absence of pointed neck and saddle hackles, giving a neat, smooth appearance all over.

The bird's head is small, with a short beak, which is dark horn-colored on Gold birds and dark blue or horn-colored on the Silver birds. The male birds have a rose comb, which should be covered with fine points. The comb narrows toward the back, with a rearward-facing leader that is angled slightly upward.

The Gold version is characterized by an even gold-bay ground color combined with sharp black lacing, which has a definite green sheen when it catches the light. The Silver version is similarly marked, but on a silver-white ground color. The legs and feet of both versions should be slate gray.

Personality

Sebrights are happy, inquisitive birds and are great to have around the garden. They also do well in mixed flocks.

Eggs

You do not keep this breed for its egg production. Laying performance seems to vary, with some owners claiming good numbers while others say they get only a few. Those eggs that are produced are small and have light-colored shells.

Day-to-day

These birds are flyers, so you will need to contain them if space is limited. They love to forage, but will also tolerate being cooped up. Breeding good ones is difficult for the inexperienced keeper because the chicks are delicate for the first few weeks and mortality rates can be high. Moreover, the hens are not known for their broodiness. Marek's disease can be a problem, and most good breeders vaccinate their stock, so check before buying.

Andalusian

attractive and graceful • flighty • reasonable layer • not for beginners

TYPE: large, light, rare • WEIGHT: large male 7–8 lb (3.2–3.6 kg), large female 5–6 lb (2.25–2.7 kg); bantam male 24–28 oz (680–790 g), bantam female 20–24 oz (570–680 g) • COLOR: Blue

The Andalusian takes its name from the province of Andalusia in Spain, and it is recognized as one of the oldest of the Mediterranean breeds. They are attractive birds, which are believed to have been developed in Britain from black and white breeding stock imported from Spain in the mid-1840s. The first attempts were more game-like in appearance and not nearly as attractive as the modern breed. Careful breeding eventually resulted in the bird we have today, with its appealing, bluish-gray plumage and black lacing. The breed is popularly referred to as the Blue Andalusian.

Looks

A good Andalusian is an attractive, graceful bird. He will adopt a bold, upright stance, standing imposingly on his long, clean, black or dark slate-colored legs. Male birds sport a fine, bright red, deeply serrated single comb, which stands proudly; on hens the comb is smaller and folded to one side. The face and wattles are red, but, in true Mediterranean style, the smooth earlobes are white.

Personality

In keeping with the breed's origins, the Andalusian has a flighty character, loves to forage, and enjoys plenty of space. Given the chance, this is a reasonable flier, and not at their happiest when being handled.

Eggs

The Andalusian is a reasonable layer of large, chalky-white eggs, and is known for an ability to often stay in production during the winter months. If the bird is in ideal conditions, you might expect about 160 eggs a year.

Day-to-day

This is probably not an ideal first bird. Their attractive appearance makes them popular among exhibitors (although they are a challenge to breed to standard), but they can be difficult for the inexperienced keeper to handle. Some owners also complain about the noise, so this might be a consideration if you have neighbors nearby. Another factor that might determine whether this is the bird for you is the Andalusian's liking for plenty of space, and overcrowding will lead to all sorts of problems, possibly even fighting. Finally, this is not a sitting breed, so they are not good for natural brooding duties. When breeding blue-to-blue, expect to hatch only 50 percent blue chicks, the rest being black or white splashes.

Successful breeding will require an incubator, and keepers will find that the chicks develop quickly. It is also worth noting that exposing the birds to excessive sunshine can turn the plumage a rusty-gray color, so a shady run is best.

Andalusian male

Asil

proud and upright • hard feather • fighting pedigree • not for beginners

TYPE: large, hard feather, rare • WEIGHT: male 4–6 lb (1.8–2.7 kg), female 3–5 lb (1.36–2.25 kg) • COLORS: Black, Dark Red, Duckwing, Grey, Light Red, Pyle, Spangle, White

The Asil, or Aseel as it is sometimes written, is a rare sight in Europe and America. It is an ancient breed—perhaps the oldest game fowl of all—and it was developed in India specifically for fighting. Its name derives from the Arabic for "long pedigree," which is particularly apt as this breed has been around for at least 2,000 years. The Asil was not bred to fight with the aid of metal spurs, as happened in Europe. Instead, it was developed for its strength alone, and it fought in battles of endurance that would sometimes last for days. As a result, the birds developed an extremely muscular form, with strong beaks, thick necks, and well-developed legs. They are, in consequence, something of an acquired taste, and it is not the easiest of breeds to live with.

Above Red Asil male, mature
Right Red Asil male, young

Looks

Like any fighter, the Asil adopts a proud, upright stance and is an active, quick mover. He stands tall, and, on a good specimen seen from the side, you could draw an imaginary line from his eye straight down to his middle toenail.

The face is covered in tough, leathery, red skin, the eyes can shut, there are no wattles, and the comb will be either a triple or a pea-type. The short, wiry feathering feels noticeably hard to the touch. There's no under-fluff below the main feathers, and patches of bare skin are visible down the front of the bird's neck and on the breast and thighs.

The coloring of this breed can vary enormously, but most often you will see Light and Dark Red forms.

Personality

Pugnacious is a word that springs to mind to describe the Asil's character. Hundreds of years of selective breeding for fighting have produced birds that are naturally aggressive with one another (even the females). However, if they are kept in isolation, they can be docile and easy to handle.

Eggs

This breed is not one of the best layers. The eggs are small and either white or lightly tinted.

Day-to-day

The Asil is a specialized rarity and certainly does not represent the best breed for the novice keeper. If necessary, they will tolerate confinement better than most other game birds, but they do love to forage. Their strength makes them very robust, durable birds and they tolerate both cold and heat well. The hens can make good, protective mothers.

Campine

flighty • showy • good layer • white eggs • attractive

TYPE: large, light, rare • WEIGHT: large male 6 lb (2.7 kg), large female 5 lb (2.25 kg); bantam male 24 oz (680 g), bantam female 20 oz (570 g) • COLORS: Gold, Silver

The Campine is an ancient European breed that was developed near Antwerp in Belgium and takes its name from that region of the country.

It was bred essentially as a utility-type egg-layer. Interestingly, the plumage color markings of the male and female birds are virtually identical. The males are what is known as hen-feathered, and they do not have pronounced sickle feathers in the tail or pointed neck or saddle hackle feathers. This interesting plumage made the Campine an important bird in early research into auto-sexing (the ability to distinguish between the sexes at a very young age through feather color differences). Much of this research was carried out in Cambridge and resulted in the creation of an auto-sexing breed called the Cambar.

Above *Note the large, prominent eyes on this young Campine hen.*

Looks

The Campine is an attractive, neat-looking bird with a compact and graceful body shape. The dark eyes are prominent, and on females the single comb falls to one side. The face and wattles are bright red; the earlobes are white. The neck feathers are pure white on the Silver version and form a cape. The rest of the body is white with pronounced beetle-green barring on all feathers. On the Gold version the pure white feathering is replaced by a rich gold color. Both types of bird stand on reasonably long, dark blue legs.

Personality

The Campine is a generally alert and lively bird, although the temperament of individual birds can vary: some will be friendly and approachable, but others can be aloof. They are great characters and very friendly, often coming to greet you as you arrive in their run.

Eggs

Primarily developed as a layer, the Campine is a reasonable producer of medium-sized, white-shelled eggs.

Day-to-day

The Campine is a flighty bird and will need a secure compound. The hens are not good sitters, but the breed is generally hardy. The only real weak point is the large comb, which can suffer from frostbite in really cold weather.

Campines are quick to feather, but take their time to reach full maturity. They're economical eaters and love to forage, but they will tolerate close confinement if necessary.

Right *Silver Campine male*

Houdan

unusual crested looks • docile • easy to handle • utility roots

TYPE: large, heavy, rare • WEIGHT: large male 7–8 lb (3.2–3.6 kg), large female 6–7 lb (2.7–3.2 kg); bantam male 24–28 oz (680–790 g) bantam female 22–26 oz (620–740 g) • COLOR: Mottled

The Houdan is an interesting French breed that was developed in a region just to the east of Paris. Its striking physical features, which include an impressive head crest and the five-toed foot, link it to the breeds from which it was created: the Poland (the crest) and the Dorking (five toes). It is available in both large and bantam forms, but this is a rare breed.

Looks

The Houdan is a distinctive bird. The rich, black feathering is evenly mottled with pure white over most areas. The body is long and deep, and the head is reasonably large and topped with an impressive crest of backward-facing feathers, which expose the leaf-type comb. Beard and whiskers cover much of the face, and the wattles and earlobes are small.

The bird stands on short, strong, featherless legs, which are light in color, with gray-blue mottling.

Personality

These are reasonably active birds, but at the same time they remain docile and easy to handle.

Eggs

Houdans are not fantastic layers. The white eggs can be a little on the small side, and owners can expect 140–160 eggs in a season.

Day-to-day

This is not the best breed to rely on if you are interested in natural brooding. It also requires quite a lot of regular attention, and the head-crest feathers suffer in wet or freezing conditions.

On the plus side, this breed will bear confinement reasonably well, is fast to mature, and is an economical eater. All in all, however, Houdans are probably not the ideal first bird for a newcomer to chickens.

Houdan female

Ixworth

very rare • attractive and neat • needs space • active

TYPE: large, heavy, rare • WEIGHT: large male 9 lb (4.1 kg), large female 7 lb (3.2 kg); bantam male 36 oz (1.02 kg), bantam female 28 oz (790 g) • COLOR: White

The Ixworth is an all-too rare sight these days. It is a traditional British breed, which has gradually slipped from favor and was nearly lost altogether in the 1970s. Thankfully, due to a few dedicated breeders, the situation is now a little healthier than it used to be.

It is a dual-purpose breed, created by Reginald Appleyard in the 1930s and named after the Suffolk village of its birth. The aim was to produce a top-quality, fast-maturing table bird that would also lay a good number of eggs, and the result seemed to fulfill all expectations.

Looks

The Ixworth is an attractive, neat-looking, pure white bird, with a bright red, pea-type comb, face, wattles, and earlobes and a light-colored beak and legs.

The body is compact but has an elongated look to it, thanks to the gentle slope of the tail. The broad head is carried on an upright neck, feathered with neat hackles.

Personality

These are active birds, which will thrive in spacious surroundings.

Eggs

A healthy Ixworth hen will produce about 150 medium-sized, light brown eggs a year.

Day-to-day

The Ixworth is best suited to a free-range life, with plenty of space. They are generally robust birds, so there should be few problems.

Because the breed remains so rare, it is important to research any birds you are intending to buy. Make sure that you are being sold actual Ixworths and not some pale imitations.

Ixworth male

Jersey Giant

large • slow maturing • good layer • fairly docile

TYPE: large, heavy, rare • WEIGHT: large male 13 lb (5.9 kg), large female 10 lb (4.55 kg); bantam male 3 3/4 lb (1.7 kg), bantam female 2 1/2 lb (1.13 kg) • COLORS: Black, Blue, White

As the name suggests, the Jersey Giant is a giant among chickens. They grow into just about the heaviest chickens you can buy, with male birds reaching a hefty 13 lb (5.9 kg), and even the hens can top the scales at 10 lb (4.5 kg).

The breed was developed in the late 1800s in New Jersey, US, as a dual-purpose bird that would be good for both meat and eggs. It is thought that that Dark Brahmas, Black Javas, Black Langshans, and Indian Game (Cornish Game) were all used in the crossing process that eventually led to the Jersey Giant.

The breed never really caught on commercially, primarily because it is a relatively slow developer—it takes about six months to get to its potential weight, which is just not quickly enough for industry needs. It has also been suggested that black-feathered birds with dark legs never earn lasting favor in the American market, and that this is another reason why the breed never fired the imagination.

Looks

Despite his size, the Jersey Giant is an alert and well-proportioned bird. He has a broad breast, long and almost horizontal back, and a full, well-spread tail, which is held at an angle of 45 degrees.

The Black Jersey Giant has lustrous black plumage, with a green sheen, a short, mainly dark beak, dark eyes, dark legs (which lighten with age), and bright red comb, wattles, face, and earlobes. Interestingly, the bottoms of the feet are yellow.

The White version is pure white all over, but with lighter, willow green beak and legs (and the yellow coloring under the feet), and the Blue has a laced pattern.

Personality

Like most large breeds, the Jersey Giant is a calm and docile bird. They are easily handled and well suited to the domestic environment.

Eggs

This is a good layer, producing about 180 brown-shelled, medium-sized eggs a year.

Day-to-day

These are big, rugged birds that take a while to mature. If you are keeping them for meat, you will need to allow six months for them to reach their full size.

The hens make good mothers, although their size and weight can cause problems with broken eggs. Their size must also be taken into consideration when you start to plan their housing. The breed will forage, but not enthusiastically, and is robust and resistant to cold.

The Giant has lots to offer the enthusiastic keeper and is a breed that deserves more attention.

Black Jersey Giants (female in foreground)

Lakenvelder

rare • excitable • unique coloring • not for beginners

TYPE: large, light, rare • WEIGHT: large male 5–6 lb (2.25–2.7 kg), large female 4 1/2 lb (2 kg); bantam male 24 oz (680 g), bantam female 18 oz (510 g) • COLOR: Black and White

The Lakenvelder is an interesting but rare breed. There is plenty of argument about where and when it originated, although most people agree that it is one of the old, established breeds.

Both Germany and the Netherlands seem to claim the credit for giving the world the Lakenvelder. The Germans say it was developed in the mid-1800s in Westfalen, in the province of North Rhine-Westphalia. On the other hand, the breed's characteristic black and white coloring is said to have been found in poultry from the Dutch village of Lakervelt, in the southeast of that country.

The name Lakenvelder has variously been translated as meaning "field of linen" or "shadow on the sheet," references that obviously relate to the bird's distinctive markings—unique among chickens. However, there is a breed of Dutch Lakenvelder cattle, which shares the same black-white-black color scheme, and this belted pattern can also be found on goats and even guinea pigs. It can be a difficult color scheme to achieve.

The Lakenvelder has never really caught on as a popular breed outside mainland Europe, which is a shame because it has plenty to offer the dedicated keeper.

Looks

The bird gives a bold impression, with his longish body, prominent breast, neatly carried wings, and jaunty tail. The smallish head is topped with a single, evenly serrated comb. Other features include a dark horn-colored beak, bright red or brown eyes, small, white earlobes, and a bright red face and wattles.

The neck is of average length with plenty of long hackle feathers, and these, like the tail, should be solid black. The legs, which are featherless and well spaced, should be slate blue in color. The rest of the body should be pure white, although some black-tipping to the saddle feathers is permitted on male birds.

Personality

In common with many other light breeds, the Lakenvelder has a generally flighty and excitable character. It does not much like human contact and so perhaps is not the ideal breed to keep in a family environment.

Eggs

The breed's laying performance is pretty good, although owners should expect the 160 or so whitish-shelled eggs that will be produced in a season to be on the small side.

Day-to-day

Although the Lakenvelder does tolerate confinement pretty well, their active and excitable nature means that enclosures will need to be secured to prevent escape. The hens do not make good broodies, but the breed as a whole is hardy and will forage happily given the opportunity.

Lakenvelder male

Malay

intimidating looks • good with people • very tall • needs space

TYPE: large, hard feather, rare • WEIGHT: large male 11 lb (5 kg), large female 9 lb (4.1 kg); bantam male 42–48 oz (1.19–1.36 kg), bantam female 36–40 oz (1.02–1.13 kg) • COLORS: Black, Black-Red, Pyle, Spangled, White

This breed, which is said to have originated in eastern India, has been kept in Britain since the early 1800s and has been used as a building block for many modern breeds. The Malay was present at the first-ever poultry show in Britain—it was held in 1845—and the breed was standardized 20 years later in its White and Black-Red forms.

The long legs and neck, coupled with an upright stance, mean that the largest examples of the Malay have been not far short of 3 feet (1 meter) tall. Despite the impressive height, it is not the prettiest of birds. Minimal feathering and a cruel-looking face (with prominent eyebrows over deep-set, light-colored eyes) give the breed a fierce, menacing appearance, which is not to everyone's taste. It is one of those breeds that people either love or hate—it is hard not to have a definite opinion about the Malay.

Looks

The main features of this striking-looking bird are a broad head, a long, wide neck, muscular, well-defined legs, and a shortish, drooping tail. Viewed from the side, a good example will present a series of curves, running down the neck, across the back and down along the tail.

The bird's head is topped by a small, walnut-like comb, which is set well forward. The comb, face, small wattles, and earlobes are all bright red.

The color variations noted above represent the standardized options, but breeders do produce other colors. However, as far as showing is concerned, the Malay's coloring is of low importance. The greatest emphasis is always placed on the bird's type—his overall appearance and bearing. Head, legs, and feathering are the other important areas of interest to the judges.

Personality

Despite his slightly intimidating appearance, the Malay can be a fairly placid bird, certainly compared with other game breeds.

Eggs

Malays are not strong layers; the medium-sized eggs are light brown.

Day-to-day

The Malay is not best suited to close confinement. This is a breed that really needs to be active if it is going to thrive, so you will need space to keep these birds. They are slow-maturing but hardy, and the females can be broody, so natural brooding is an option.

Above *Black-Red Malay female, young*
Right *Black-Red Malay male*

Marsh Daisy

hardy • slow to mature • tinted eggs • calm nature

TYPE: large, light, rare • WEIGHT: male 5 1/2–6 1/2 lb (2.5–2.95 kg), female 4 1/2–5 1/2 lb (2–2.5 kg) • COLORS: Black, Brown, Buff, Wheaten, White

The Marsh Daisy is a rarity these days, but well worth owning if you can track down a reputable breeder. The breed arose from a complicated sequence of crossings that took place in Britain over a 35-year period beginning in the 1880s in Southport, Lancashire. It all started with an Old English Game bantam cock, which was crossed with Malay hens. Then Hamburgh and Leghorn blood was subsequently introduced to the mix, and this was followed by various other game breed crosses and, finally, the attractively named Sicilian Buttercup.

Brown Marsh Daisy male (right) and females

Regrettably, such is the rarity of the breed that it is unlikely that you will come across any of the Black, Buff, or White forms. Only the Wheaten and Brown varieties are left. There is no bantam version of this breed.

Looks

They are attractive, traditional-looking chickens that grow to a good size, with a plump breast and a broadness at the shoulder. The head is dominated by an impressive rose comb, which is evenly spiked and has a single leader that protrudes out toward the back. The short, curved beak is horn-colored, and the prominent eyes are red, as are the face and wattles. The earlobes are often a combination of red and white, but always more red than white.

The neck is reasonably long, with attractive hackle feathers that fall down to form a cape around the shoulders. The bird stands on good length, featherless, willow-green legs, with four toes on each foot. The Brown and Wheaten varieties provide attractive combinations of rich gold, black, and brown coloring, and both are appealing.

Personality

The Marsh Daisy combines a bold, active nature with a generally calm character, although, they can fly easily if disturbed.

This Marsh Daisy male has richly colored hackle feathers.

Eggs

This is a reasonably good layer of medium-sized, white-shelled eggs, although some would argue that the eggs are not as big as they used to be.

Day-to-day

The Marsh Daisy is a good forager and a generally hardy bird. They will get used to confinement if they have to, but prefer to have more space rather than less. The breed is slow to mature and is an economical eater. Also worthy of note is the fact that the Marsh Daisy will endure wetter conditions than most other breeds.

Unfortunately, the breed's rarity means that sourcing good stock can be a problem; established breeders are few and far between.

New Hampshire Red

homely breed • robust • good mother • great in the backyard

TYPE: large, heavy, rare • WEIGHT: large male 8 1/2 lb (3.85 kg), large female 6 1/2 lb (2.95 kg); bantam male 34 oz (960 g), bantam female 26 oz (740 g) • COLOR: Red-Brown

The New Hampshire Red was developed as a dual-purpose breed directly from the Rhode Island Red. It appears that farmers in New Hampshire wanted to emulate their counterparts in Rhode Island by having their own distinct breed of chicken named after their state, but rather than set about the task with a complicated program of inter-breed crossing they simply spent 30 years refining the Rhode Island Red.

It was a slow process, and selection was based purely on utility grounds. The aim was to create a bird that was a good and rapid meat producer, as well as being a fine layer of large eggs. The end of the process arrived in 1935, when the breed was standardized in America, at which point the first examples started appearing in Britain.

Despite its formidable laying performance—trials in America recorded one bird producing 332 eggs in a 52-week period—the breed was not an immediate success in Britain, where it languished for many years in the shadow of the better established and popular Rhode Island Red breed. It was probably the creation of the bantam version in the 1980s that led to the breed's increased popularity with poultry keepers. It is a breed that has a great deal to offer today's poultry keeper.

Looks

The original creators of the New Hampshire Red never claimed that it was a great "looker," but it is now widely regarded by keepers as an attractive and homely breed. The well-rounded, deep body gives

New Hampshire Red female

the bird a traditional appearance, and the head, which tends to have a flattish top, features a proud, single red comb with five, well-defined points. The eyes are prominent and set quite high. The bird's face is smooth-skinned, and the wattles and oval-shaped red earlobes are moderately sized. The head sits on a well-proportioned neck covered in flowing hackle feathers, and the bird's rich yellow legs are widely spaced and strong, with four toes on each foot.

The plumage is a rich mixture of reds, browns, and golds. The male birds have black tail feathers and also some black edging on the wing feathers. The females show black tipping to the feathers on the lower neck and dark edging on the wing and tail feathers.

Personality

This American breed has a great personality and is ideally suited to a backyard. It has been suggested that some strains can be aggressive, but this is rare. In general, the New Hampshire Red is a friendly, docile bird. He will reward his keeper with a generally friendly attitude and will happily scratch around in a domestic environment.

Eggs

These days the typical performance of hens from this breed is well down on the results achieved in the 1930s. Owners should expect a New Hampshire Red to produce about 140 eggs in a season.

Day-to-day

These are good backyard birds. As long as they are not overcrowded, they are happy to be confined, but will also be content to free-range and forage if they are given the freedom to do so. It is generally robust and hardy, although the combs can be vulnerable to frost. The hens make good broodies and caring, attentive mothers.

Remember that, being a large, heavy breed that is not terribly active, the New Hampshire Red can suffer from overfeeding.

New Hampshire Red male

Norfolk Grey

unusual • docile • friendly character • good mother

TYPE: large, heavy, rare • WEIGHT: large male 7–8 lb (3.2–3.6 kg), large female 5–6 lb (2.25–2.7 kg); bantam male 32 oz (910 g), bantam female 24 oz (680 g) • COLORS: Silver-White and Black

This breed was developed in Britain by a Norwich-based breeder called Fred Myhill and first came to prominence at a dairy show held in 1920. Unfortunately, this bird, which was developed as a utility breed, never really caught on and 1970s stocks reportedly dwindled to just four birds—one male and three females. Thankfully, these birds fell into the hands of sympathetic enthusiasts and the breed was saved. It is still classified as rare, but the breed's future seems assured.

Norfolk Grey male

Looks

The Norfolk Grey has a long body with broad shoulders, a full breast, and a well-feathered tail. The head has large eyes and a single and serrated comb. The wattles are long, but the earlobes are small. The head sits on a well-hackled neck. The dark legs have four-toed feet.

Male birds have a profusion of silver-white feathering on the neck (with black striping), back, and saddle, and the rest of the bird is black. The hen is similar but with some silver lacing on the breast.

Personality

This bird has a mainly docile and friendly character and can be an endearing companion in the garden.

Eggs

Norfolk Grey hens will lay a reasonable number of medium-sized, light brown eggs.

Day-to-day

This hardy breed loves to forage, but is perfectly happy in confinement if necessary. The hens can go broody and make good mothers.

North Holland Blue

ideal family bird • docile • great layer

TYPE: large, heavy, rare • WEIGHT: large male 8½–10½ lb (3.8–4.8 kg), large female 7–9 lb (3.2–4.1 kg); bantam male 42 oz (1.19 kg), bantam female 36 oz (1.02 kg) • COLOR: Blue-Grey Cuckoo

This Dutch breed is a good dual-purpose choice for those seeking a regular supply of eggs and a tasty roast bird every now and then.

The North Holland Blue was created to meet the increasing demand for eggs in the late 19th and early 20th century. Breeders used Belgian Malines crossed with others, including Plymouth Rocks, Sussex, Orpingtons, and Rhode Island Reds.

Looks
This is an attractive breed with its all-over cuckoo plumage pattern, which contrasts well with its red comb, lobes, wattles, and face. The body is compact and neat, with an upright, alert stance. The tail is well spread and carried at an angle of about 45 degrees. A short beak and bold red-orange eyes give the bird a keen expression. The single comb is evenly serrated with between five and seven points, and the earlobes and wattles are medium sized. The light-colored legs are widely spaced and lightly feathered, according to the British standard. In the Netherlands, however, they are featherless.

Personality
The breed exhibits an almost ideal combination of a quiet, docile nature with practical, utility qualities. They are great family birds.

North Holland Blue male

Eggs
A young and healthy hen will lay about 180 eggs a season, although some will lay closer to 240 eggs.

Day-to-day
This is a practical breed for keepers who want attractive birds that are also productive. The birds love to forage and are ideal for free-range living. If they are confined, take care not to overfeed them because they can easily become overweight.

Orloff

Russian origins • bearded • hardy • reasonable layer • very rare

TYPE: large, heavy, rare • WEIGHT: large male 8 lb (3.6 kg), large female 6 lb (2.7 kg); bantam male 40 oz (1.13 kg), bantam female 36 oz (1.02 kg) • COLORS: Black, Cuckoo, Mahogany, Spangled, White

The Orloff is a rare breed. It originated in northern Iran, from where it went to Russia, where it was named by Count Orloff Techesmensky. This happened in the 1880s, and since then the Orloff has gradually spread around the world. The first dedicated breed club in Britain was established in the 1920s, and in 1925 the bantam version of the breed first appeared in Germany.

In America, where the breed is simply known as the Russian, it is most commonly found with black feathering. It is, however, no longer listed in the American book of poultry standards.

The first birds were Malay-like in appearance, but at some point the breed was crossed to produce more utility characteristics.

The Orloff displays a full beard and whiskers.

Looks

The Orloff has a slightly odd, almost top-heavy appearance. The thickly feathered neck, beard, and whiskers make the head appear small, an impression enhanced by the small, walnut-type comb. The small wattles and earlobes are often hidden by feathers. The eyes are shaded by heavy brows, which can give the bird a rather gloomy expression. The bird is upright with a longish, sloping body. He stands on reasonably short legs with four well-spread toes on each foot.

The Black form is solid black to the skin, with a beetle-green sheen. The Cuckoo has a light blue or gray ground, with each feather darkly banded. The Mahogany is a mixture of dark browns, orange, and black, and the Spangled has orange hackles with white tips and black and white feathers elsewhere. The White has lustrous white plumage from head to tail. In all forms, the beak and legs should be yellow.

Personality

The breed has a reputation for being calm but not docile.These birds are not particularly happy being handled.

Eggs

The Orloff is a reasonable layer of smallish, tinted eggs. Hens in good health and in the right conditions should produce about 150 eggs a season.

Day-to-day

This is an appealing and interesting breed, but it will not be easy to find a source of good birds. The beard and whiskers need regular inspection to avoid problems. They are hardy foragers, who will also tolerate confinement if necessary, and they will coexist quite happily with other breeds.

Spangled Orloff male

Shamo

tall • menacing • territorial • friendly • good mother

TYPE: large, hard feather, rare • WEIGHT: male 7 1/2 lb (3.4 kg), female 5 1/2 lb (2.5 kg) • COLORS: Assorted

Wheaten Shamo female

The Shamo is a an exotic breed that can be expensive to buy but is fascinating to own. These birds can be troublesome, however, not so much because of problems with the birds themselves, but because of the attention they can attract. They were created as fighting birds and are still used as such by some owners. Consequently, good examples have a considerable black market value, and thefts—from both private premises and poultry shows—are unfortunately common.

The breed was developed in Japan after being taken there from Thailand in the 17th century. The Japanese turned the Shamo into a formidable fighting bird, and they became famous for the highest levels of ferocity and courage.

In Japan the breed is subdivided into two size types—the large O-Shamo and the smaller Chu-Shamo—but in Britain this distinction is not officially recognized and both types are bundled together under the single Shamo label.

The Shamo has an almost prehistoric appearance, and there can be no denying the power and purpose of their muscular, upright bodies. It is a breed that is hard to be indifferent about: most people either love it or hate it.

Looks

The Shamo male has a fierce, dominant appearance, and even the female looks aggressive. Long, powerful yellow legs support a sleek, well-muscled body that is held almost upright in the male.

The neck is thick, strong, and slightly curved, leading up to the head, which has a broad, light-colored beak, cruel orange eyes, and tiny wattles and earlobes (if any at all). The comb is compact and bright red.

Other notable—and slightly scary—features include patches of bare red skin running down the keel, at the throat, and on the tops of the wings. These help to intensify the rather menacing air that surrounds this formidable game breed.

Personality

Despite his aggressive appearance, the Shamo can actually be a pleasant bird to own as long as he is kept in the right conditions and he is reasonably even-tempered.

Eggs

Birds of this sort are never going to compete with the established laying breeds and should not be expected to. The shells range from white to tinted.

Day-to-day

The Shamo's aggressive streak will show itself if you put males together in the same pen or with other breeds, so take care. There have been reports that even the chicks will fight each other, so you will need space, careful segregation, and plenty of good husbandry to get the best from this breed. Make sure that your property security is tight.

A large Shamo male dwarfs young females.

Sicilian Buttercup

unusual comb • good layer • excitable • not for handling

TYPE: large, light, rare • WEIGHT: large male 6 1/2 lb (2.95 kg), large female 5 1/2 lb (2.5 kg); bantam male 26 oz (740 g), bantam female 22 oz (620 g) • COLORS: Gold, Silver

The Sicilian Buttercup is a distinctive bird famous for the unusual shape of his comb, although likening it to an actual buttercup is perhaps a little fanciful. It is certainly cup-shaped, and the top edge is serrated to form a ring of points, rather like a crown. It is not as pronounced on female birds. Nevertheless, the originators of the breed evidently thought that the likeness was strong enough to warrant the use of the wildflower's name as part of the breed name. The Sicilian part refers to the breed's generally accepted birthplace, although there have been suggestions that it actually came from further afield, perhaps Tripoli in Lebanon.

Whatever the truth about its origins, it is likely that the Sicilian Buttercup arose following crosses made between native Italian breeds and others, such as the Egyptian Fayoumi or even the French Houdan. Examples of the new breed were shipped to America in the mid-1830s and a Buttercup Club was formed there in 1912.

At about the same time, the first Buttercups arrived in Britain from America, having been imported by a Mrs. Colbeck from Yorkshire. Things went well initially, and by the 1960s the breed had become quite popular in laying trials. But numbers dwindled again, and nowadays this unusual bird is an extremely rare sight.

Looks

This is an active, fidgety breed. The body is long and deep (broad

Gold Sicilian Buttercup female

at the shoulder), and the long wings are kept closely folded. The tail should be held at an angle of 45 degrees, and on the males it features curved sickle feathers and plenty of coverts.

The bird's head is, of course, dominated by the crown-like comb, but he also features large red-brown eyes, almond-shaped earlobes (red and white), and thin but rounded red wattles. The neck is quite long, with plenty of hackle feathering falling down over the shoulders. The legs and feet should be free from feathers and willow green in color.

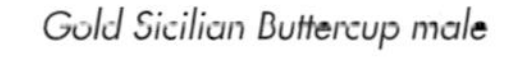

Gold Sicilian Buttercup male

Personality

In true Mediterranean style, the Sicilian Buttercup has an edgy, excitable character.

Eggs

A light breed, the Buttercup is a good layer of smallish, white-shelled eggs, and a good, healthy hen should produce 180 or more in a year.

Day-to-day

This breed is an economical eater, but its flighty temperament means that it is not well suited to human contact. Problems can also arise if there is insufficient space for the number of birds being kept. The Sicilian Buttercup loves to forage given the chance, but keepers should remember that the combs can suffer from frostbite. The hens do not make good broodies, and fencing will need to be secure and high to keep these birds where you want them.

The Sicilian Buttercup is not a breed to be kept by the novice keeper. The bird's nervous character means that dealing with him on a day-to-day basis can be awkward if you do not know what you are doing.

The fact that this bird will actively avoid contact with his owner should give you a clue about the potential difficulties associtated with this breed. If you have children in the household who will want to handle and pet the chickens, the Sicilian Buttercup really shouldn't be on your breed shortlist.

Spanish

ancient breed • needs space • flighty • noisy • white eggs

TYPE: large, light, rare • WEIGHT: large male 7 lb (3.2 kg), large female 6 lb (2.7 kg); bantam male 38 oz (1.08 kg), bantam female 32 oz (910 g) • COLOR: Black

The White-Faced Black Spanish sits at the head of the extensive group of Mediterranean birds, which includes the Ancona, Andalusian, Leghorn, and Minorca. These breeds are known as nonsitters—that is, the hens rarely settle to incubate their own eggs—which, of course, goes against natural survival instincts. To breed this out of a bird in favor of out-and-out egg production takes many generations of domestication and careful selection.

Looks

The Spanish has a longish body that slopes down toward the tail, which is reasonably well spread, but is held at a fairly shallow angle. The head, which sits on a long, well-hackled neck, is crowned by a single, serrated comb. The comb is vertical on the male, but folded to one side on the female. The face is white, as are the finely textured earlobes. The pendulous wattles are bright red and the eyes are black. The bird stands on long, thin legs, which are featherless.

Feathering should be pure black, with a good beetle-green sheen when caught in a favorable light.

Personality

Although the Spanish breed has striking looks and good laying performance, it is a flighty, excitable and often noisy bird.

Spanish male

Eggs

A hen should produce 180 large, white-shelled eggs in a season.

Day-to-day

If you want a streetwise, independent breed, then the Spanish is for you. Females are long-lived, though males rarely last more than two years. Males will need shelter in the winter due to their large combs and white faces.

Sultan

feature-packed • high maintenance • difficult to breed • friendly

TYPE: large, light, rare • WEIGHT: large male 6 lb (2.7 kg), large female 4 1/2 lb (2 kg); bantam male 24–28 oz (680–790 g), bantam female 18–24 oz (510–680 g) • COLOR: White

The breed originated in Turkey, where it was reportedly kept on the grounds of a palace by one of the sultans of Constantinople. It found its way to Britain in 1854, when a small group of birds was imported by Miss Elizabeth Watts from Hampstead, London. It has remained a rare sight more or less ever since. This is a shame because in addition to being a remarkable-looking bird, the Sultan has an appealing and pleasant character.

Looks

The Sultan is an active bird, whose body shape is disguised by profuse feathering. The bird has a deep breast, a short, flat back, large wings, and a long tail. The head is dominated by a large, globular crest and muffling (beard and whiskers). The male's small, red, V-type (horned) comb is virtually covered by the crest, but there are large nostrils that rise above the beak on each side. Below the red face is a full beard, which joins into whiskers on each side.

The legs are heavily feathered and have vulture hocks. These are large, stiff feathers that grow from the hock joint and face backward and down toward the ground. The light-colored legs are also feathered and each foot has five feathered toes.

Personality

The Sultan is a friendly bird, which when given suitable care and attention and a good living environment, will become tame.

Eggs

Historical reports suggested that they lay well, but modern literature is not as encouraging. The eggs have white shells.

Day-to-day

This nonaggressive breed would make a good family chicken, although the plumage requires constant care and attention. Narrow-lipped drinkers should be used to prevent the head feathering from getting wet, and the feathered legs and feet mean that dry conditions are essential. The Sultan is not suited to a mixed flock, because the birds could be bullied by other, noncrested breeds.

Sultan male

Sumatra

exotic looks • not a good mixer • fighting heritage • needs space

TYPE: large, light, rare • WEIGHT: large male 5–6 lb (2.25–2.7 kg), large female 4–5 lb (1.8–2.25 kg); bantam male 26 oz (740 g), bantam female 22 oz (620 g) • COLORS: Black, Blue

This elegant bird comes from the island of Sumatra, which is part of Indonesia and is situated in the Indian Ocean, close to Malaysia. He was used there essentially as a fighting bird (his legs feature double spurs) and was taken to America in the late 1840s. The breed was known by various names at this time—including the Pheasant Malay—and is believed to have been crossed with the Indian Game (Cornish Game). Nothing much more was heard about the breed until some American show-winning examples were taken to Britain in 1902 by Frederic Eaton from Norwich.

Despite the breed's indisputable beauty, the Sumatra never quite managed to fire the imagination of fanciers at large. One theory for this relative lack of interest is that by the time the Sumatra arrived in Britain cock fighting had already been banned for 50 years or so.

A standard for the breed was established in Britain in 1906, when it was known as the Black Sumatra, and a bantam strain was created as recently as the late 1970s.

Looks

The Sumatra has a regal, pheasant-like look, with the male bird being particularly notable for his long, flowing tail feathers.

The body is long, firm, and muscular. The wings are large and strong, and the shoulders are broad. The male bird's tail tends to droop, creating an elongated impression, and the long, curved sickle feathers often brush the ground. The head is small, with a strong, black beak and large black or dark brown eyes. The comb is of the pea type and, in common with the face and earlobes, is very dark (ideally black), but there are no wattles. The neck is long and well-hackled, particularly on the male. The strong legs are dark (preferably black) but not very long. The feet are a matching color, with four well-spaced toes.

Personality

The breed's fighting heritage gives the Sumatra a fearless character, although the all-black appearance does tend to make them look slightly more menacing than they really are.

Eggs

A Sumatra hen can be expected to lay about 120 light-colored eggs in a season.

Day-to-day

The Sumatra is a hardy bird that will tolerate both high and low temperatures well. They love to forage and are not keen on close confinement. Moreover, given their background, they are probably not birds that you should plan to keep in a mixed flock. The hens make good broodies.

Black Sumatra male

Transylvanian Naked Neck

bizarre looks • surprisingly practical • friendly • reasonable layer

TYPE: large, heavy, rare • WEIGHT: male 7–8 lb (3.2–3.6 kg), female 5 1/2–6 1/2 lb (2.5–2.95 kg) • COLORS: Blue, Buff, Cuckoo, Red, White

In terms of shock value, the Transylvanian Naked Neck certainly takes some beating; some viewers might even consider it alarming. Most observers have to admit that this breed is rather gruesome and ugly. To see a bird wandering around with a normally feathered body and head but a completely bare neck is shocking enough. But when the skin of the neck has the redness of a recently healed wound, it gives the impression that something has gone terribly wrong.

But this is how the breed is meant to be. It appears to be a naturally occurring freak of nature—but one that is a robust survivor. The breed is actually very old and is said to have originated, as its name

White Transylvanian Naked Neck bantam female

suggests, in the Transylvanian region of Hungary (now part of Romania). The first examples arrived in Britain from Austria as early as 1874, but little more is known about their history. It is said that some time after they had arrived in Britain a newspaper ran a story suggesting that the Naked Neck was, in fact, the result of a cross made between a chicken and a turkey.

These are hardy creatures, and they have considerable utility merit. They lay impressively sized eggs at a good rate and are also an excellent table fowl.

Looks

This breed has a large body and an alert, upright manner. The wings are medium sized, the breast is broad and well rounded, and the tail tends to be carried at a fairly low angle. The head has a single, deeply serrated red comb, large orange eyes, and bright red wattles and earlobes. The naked neck is a similar bright red, and the skin should be smooth and free from wrinkles. An interesting point worth noting is that you will sometimes find a small clump of feathers at the base of the neck, just above where it joins the breast.

The bird's legs vary in color depending on plumage. They are yellow or horn-colored on Black or Cuckoo varieties, or yellow or white on White birds.

Black Transylvanian Naked Neck male, large

Personality

These birds are easy to handle and friendly. Their generally calm and docile temperament makes them easy to tame, and they are extremely suitable for the domestic environment.

Eggs

This breed remains a reasonably good layer, and owners can expect 150 or more light-brown eggs a season.

Day-to-day

The utility strengths offered by the breed make it an attractive proposition from a practical (if not an aesthetic) point of view, with the naked neck feature widely used in modern day commercial table-bird production. These birds are economical eaters and will happily forage on a free-range basis, but they will also cope with confinement well if necessary. They make great mothers.

Vorwerk

adaptable • good layer • dual-purpose • rare • attractive

TYPE: large, light, rare • WEIGHT: large male $5^1/2$–7 lb (2.5–3.2 kg), large female $4^1/2$–$5^1/2$ lb (2–2.5 kg); bantam male 32 oz (910 g), bantam female 24 oz (680 g) • COLORS: Buff and Black

One of the Vorwerk's claims to fame, apart from its unique coloring and useful, dual-purpose potential, is the fact that it is the only breed of chicken that shares its name with a vacuum cleaner (made in Germany, of course).

The Vorwerk was developed in Hamburg by German poultry breeder Oskar Vorwerk. In 1900, he set out to create a genuinely practical, medium-sized utility fowl. Not only did he want the bird to be a good egg and meat producer, but he also aimed to produce a chicken that was an economical eater, mild-mannered, and easy to keep. He succeeded on all counts.

It is thought that much of Oskar Vorwerk's thinking was based on the idea of creating a dark buff-colored version of the Lakenvelder breed, with a similar belted color pattern. Popular belief has it that he preferred the darker ground color because it meant that the bird was less inclined to show dirt.

Consequently, it seems likely that the Lakenvelder was used as the starting point in the breeding process and that other breeds involved included the Orpington and the Andalusian. Unfortunately for Vorwerk, his creation never really gained universal acceptance, even though it was standardized in 1913.

Today, the Vorwerk is confined to a few enthusiast keepers who appreciate its many qualities and

Vorwerk female

preserve its heritage. Numbers are so small that there is no specific breed club to support its needs, although, thankfully, in Britain the Rare Poultry Society does include the Vorwerk under its protective rare breed umbrella.

Looks

The Vorwerk has a typical utility shape: a broad, deep body, a well-rounded and full breast, and a tail held at a lowish angle. The medium-sized head has an average single comb that has up to six serrations. The red face is covered with tiny feathers, and the bird has alert orange-red eyes, medium-sized wattles, and small, white earlobes.

The upright neck is covered with full hackle feathers, and the bird stands on slate gray, featherless legs with four-toed feet. The plumage color scheme is similar to a Lakenvelder, with the white ground color replaced by an attractive dark buff. The neck and tail should both be pure black, and the feathers, generally, are fairly close-fitting.

Black spotting in the buff-colored areas is a bad point, although consistently well-marked birds are difficult to breed.

Personality

These alert and active birds have a very suitable character for the domestic keeper.

The Vorwerk is a great dual-purpose breed; males birds like this will be good for the table.

Eggs

A well-bred Vorwerk hen should produce an acceptable 170 cream or tinted eggs in a season.

Day-to-day

Vorwerks are adaptable birds that will do well under almost any conditions, as long as they receive good levels of care and attention. However, they can fly, so you will need to take the necessary precautions to keep them contained. They are happy on small amounts of good-quality feed, and the chicks are robust and quick to grow.

For keepers interested in growing the occasional bird for the table, the Vorwerk is suitable for this purpose, too.

Yokohama

spectacular long feathers • good mother • aggressive males

TYPE: large, light, rare • WEIGHT: large male 4–6 lb (1.8–2.7 kg), large female 2 1/2–4 lb (1.1–1.8 kg); bantam male 20–24 oz (570–680 g), bantam female 16–20 oz (490–570 g) • COLORS: Golden or Black-Red, Golden Duckwing, Red-Saddled, Silver Duckwing, White

The Yokohama, as it is known in the West, is essentially a Japanese breed, although its roots can be traced back to ancient China. Japanese research suggests that it was created by crossing the Shamo with a long-tailed Chinese breed called the Shokoku. It is not, of course, called the Yokohama in Japan; this was simply a name given to it at some point after the breed arrived in the West, most probably in reference to the port from which the early shipments were dispatched. The first examples arrived in Germany in the 1870s.

Red-Saddled Yokohama female, young

Because it is a game-type breed, it is likely that these birds, and their long-tailed relations, were used for fighting in Japan, albeit as part of religious ceremonies. The closest Japanese version to the European Yokohama is called the Minohiki, although it is difficult to be precise because there are a number of other breeds with similar names.

The birds are highly revered in Japan, where the males are characterized by exaggeratedly long tail and saddle feathers. The sickle feathers grow at a tremendous rate and molt out only every three years or so. The longest recorded was well over 20 feet (6 meters). In Japan, the birds are kept on high perches and fed a special diet to promote tail growth. They are carefully exercised for a short period each day, during which the tail is supported to prevent damage.

In the West, the tails are unlikely to grow to these extraordinary lengths because the environmental conditions are different, and the birds will molt on a conventional yearly basis.

Looks

The Yokohama is a graceful-looking bird. The males, in particular, with their flowing saddle and tail feathers, are striking.

The long body has a full, rounded breast and is carried rather like that of a pheasant. The tail is held fairly low. A small head sits on top of a long, well-hackled neck, and there is a small single, walnut or pea-type comb. The earlobes and wattles are small also, and everything is red, including the face; the earlobes can be white on Black-Red and Duckwing birds. The eyes are bright, lively, and red, and the yellow to horn-colored beak is strong and curved. The bird stands on clean legs and four-toed feet, which may be yellow, willow green, or gray-blue.

The five plumage color varieties noted above are those standardized in Britain, but others do occur. In showing circles, the greatest emphasis is placed on the bird's overall type and on the tail and saddle feathers; color is far less important.

The Yokohama certainly presents a unique image, offering a stately presence for any backyard.

Personality

The Yokohama is generally calm and gentle, although keepers should bear in mind that males can be aggressive with each other.

Eggs

These birds are not the good layers they once were. You should expect 80–100 tinted or white eggs in a season.

Day-to-day

The Yokohama's long feathers mean that it is not the easiest breed to keep in a restricted environment, and it is essential that damp conditions are avoided. The birds are relatively slow to mature, but will happily forage if allowed to do so. The hens make very good mothers, so natural incubation is a possibility.

Red-Saddled Yokohama male

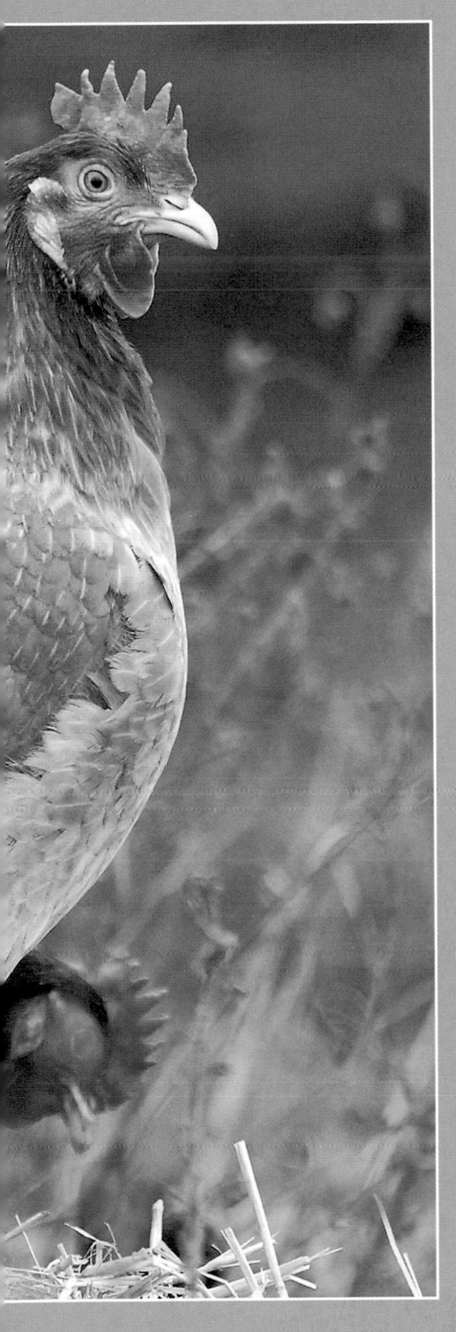

Hybrids—The Modern Alternative?

If you eat a lot of eggs and want to keep chickens that are certain to provide your family with a healthy, fresh supply of one of nature's best all-in-one foods, you should think carefully about the hybrid option.

Eggs galore

Established pure breeds are the birds that dominate both domestic poultry keeping and the exhibition world. But if you want to guarantee high egg production, the hybrid hens are the answer.

The choice among pure breeds, in terms of plumage color, size, and behavioral characteristics, is enormous, as a glance through the preceding chapter will reveal. Moreover, the fact that these breeds have been around for years and years means that, for many people, there is a historical aspect to the breed they pick. Many keepers, for example, can remember parents or grandparents owning a specific breed, and they have an emotional attachment to it for that reason alone.

Then there is the conservation issue to consider. The idea that you are doing a genuinely worthwhile thing by selectively breeding and carefully maintaining good, representative stock is a deeply satisfying one. Some of the more obscure pure breeds now exist in worryingly small numbers, and they will not be saved without the dedicated efforts of enthusiastic owners who are prepared to put in the time and effort needed to ensure their survival.

Survival of the fittest

Having said all that, the converse of the argument is that, by and large, breeds of any living thing tend to die out because they are lacking in some way. Apart from those creatures that humans have hunted or polluted into extinction, breeds generally come and go in response to the natural environment around them. Unfortunately for chickens—and all other species that have been domesticated—their survival depends on we humans and our whims. Cold-hearted logic dictates, therefore, that those breeds of poultry that now find themselves teetering on the brink of oblivion are in that position only because of their own shortcomings.

Remember that human intervention led to the creation of virtually all the chicken breeds we have today and that these chickens are the result of a long process of careful and imaginative crossbreeding. Some of the results have proved to be extremely successful, but others have not, and it is the members of this latter group that now find themselves in decidedly choppy water.

In many respects, the poultry exhibition scene has a lot to answer for in terms of the effects that generations of show breeders have had on many of the most popular pure breeds. The desire to breed and show larger and more profusely feathered birds often came at the expense of egg-laying ability. The fact that both

Above *Laying performance varies enormously from breed to breed.*

Above *A fresh egg for breakfast! Hybrid hens are the best layers you can get, and a good one will produce more than 350 eggs a year.*

eggs and feathers have a high protein content means that a chicken can effectively produce only one or the other. This is why hens stop laying eggs during the molt—all their energy (and protein) is going into the production of new feathers, and there is nothing left for the egg side of the operation. Breeds like the Cochin, which arrived in the West to great acclaim, thanks to their large brown eggs and winter-laying ability, were successively bred for more and more feather. As a consequence, the best of their utility function was lost, and today's birds do not perform nearly as well in the nest box as their forebears.

It goes without saying that the commercial poultry industry has never been slow to appreciate the difficulty of balancing appearance and utility. It has always had a vested interest in maximizing the performance of the breeds it uses for both egg and meat production. More recently, though, specialist breeders have started developing a range of high-performance yet colorful hens—the best of both worlds?

Commercial interests

Much work was done, initially in the United States, that highlighted the differences in laying performance among the breeds. The effects on egg production, in particular, that resulted from crossbreeding were also carefully recorded.

Early crossbreeds

After the First World War ended the demand for fresh eggs increased dramatically, so selective breeding programs using established high performers, such as the Rhode Island Red, became increasingly important.

Up to the Second World War, it was quite common for poultry farmers in the United States to be producing 250 eggs from each bird in a year—impressive figures, indeed. Expert breeders took great care when they were crossing strain with strain, and when necessary they introduced fresh blood from other, high-performing utility breeds, such as the Light Sussex. And, in this way, they created what was to become known as the modern hybrid chicken.

Above *This is a pretty Marans-based hybrid called the Speckledy. These hens will lay large numbers of dark brown, speckled eggs.*

The influence of the hybrid has been significant. Essentially, there are two types: the egg-layer and the meat bird. These have been the mainstay of the developed world's poultry industry for the past 50 years, and they have transformed chicken from what was formerly a luxury food to an affordable, everyday commodity. Whether this is a good thing is another matter altogether. The intensive methods used in the modern industry certainly are not to everyone's taste. Birds that are reared for their meat have a life span of little more than six weeks, and they are treated with all sorts of chemical growth promoters to ensure that they reach the required weight in the shortest possible time. It is a profit-motivated industry and, as long as the majority of consumers continue to buy on price alone, these harsh, high-pressure conditions are likely to remain.

Hybrid layers

Interestingly, increasing numbers of hybrid breeders are starting to make stock available to the hobby keeper.

The possible benefits of hybrids compared with the traditional pure breeds to hobby keepers are quite significant. Hybrids are usually a good deal cheaper to buy, they are supplied effectively vaccinated against the most serious poultry diseases, and, of course, they will lay like there's no tomorrow.

On the downside, some people consider them rather uninteresting to look at. There is certainly not the variety of plumage coloring that is available when you buy pure breeds, and hybrid birds can be rather unpredictable if you try to breed from them, which is not true of good-quality, pure-bred stock.

Above *You could not ask for a better range of egg colors than these from the Fenton Blue hybrid hen.*

For the beginner, however, hybrids can represent a sensible, even-tempered, docile choice, and their tremendous laying ability is bound to please everyone, with the best examples laying more than 350 eggs in a season.

Even though the choices available are not as varied as with pure breeds, you should not imagine that there are only two or three to choose from. Nowadays, there are plenty, and the number of options is growing all the time. Most are hybrids based on the Rhode Island Red, and so they will lay light brown eggs, but you can also get Leghorn-based versions if you prefer, which will lay brilliant white-shelled eggs. For something even more exotic, you should look out for the newly created hybrid, originally developed in Britain, which is called the Fenton Blue and which lays an assortment of tinted eggs ranging in color from olive green to pale blue.

Meaty options

Increasing numbers of backyard keepers are starting to keep a few birds for their meat. Now, although this is certainly not going to attract everyone, the idea of rearing your own birds under the best possible conditions, free from chemical additives and growth enhancers, is an increasingly appealing one. Some meat-eaters are starting to question exactly what goes into the supermarket product. Consequently, many people feel happier eating and feeding their children on birds that have led contented, healthy lives and that have been humanely killed.

You can eat just about any breed of chicken, but some are better than others. The hybrid option, of course, opens the door to a range of much more specialized breeds that have been developed specifically to grow quickly. These birds are known collectively as "broilers," an American term that was coined to describe a fast-growing young chicken that is grown for meat. Nowadays, the mass production of chicken meat is an enormous global industry, dominated by three or four companies, which supply around 90 percent of the world's breeding broiler stock. In the European Union alone more than 4.5 billion chickens are produced annually, and in Britain consumption is about 66 lb (30 kg) per person per year.

The practical ins and outs of the modern broiler industry don't make happy reading and will not be described here. However, the point worth making is that it is now possible for the domestic keeper to purchase from a range of "meat birds" that, with a little care and attention, can be reared for the table perfectly happily in the domestic environment. Those keepers who are interested can grow everything from a small poussin to a large roasting chicken.

Traditionally, one of the problems associated with rearing these specialized birds in the domestic setting was that they are easily overfed, which can cause heart problems and place undue stress on the legs—the resulting weight gain is simply too much for the bird to support. In a bid to counter these unfortunate problems, breeders are now starting to select lines that still give good all-around performance but with a slightly slower growth rate. These, in combination with the use of lower protein feed, can make all the difference.

Rescue mission

One final hybrid option that is gaining momentum in some countries is that of battery hen rescue. Laying hens are saved from slaughter by caring keepers offering a restful and stress-free retirement.

Double standards

The whole issue of battery hen farming is an emotive one. In fact, many consumers prefer not to think about the conditions endured by the birds during their short, 72-week lives. The whole business is beset with rules and regulations, argument and protest, and a fair amount of double standards.

On the one side, farmers and producers say that they are working within the raft of animal welfare laws that regulate the industry and that they are simply meeting the public demand for cheap chicken eggs and meat. Opposing these views are those who consider battery farming to be obscene and cruel, arguing that keeping a hen in a wire cage with a floor area no bigger than a sheet of paper—there is no room for the bird to turn around—is little short of barbaric.

Above *Ex-battery hens are not a pretty sight in the early stages, but most recover to enjoy their well-earned retirement.*

Then there are those in the middle, who, without actually supporting the industry, maintain that the birds in a typical battery unit are actually pretty healthy (sick birds do not produce eggs, and as such are of no use to the farmer) and have never known anything better than their caged existence. Finally, there are members of the public who think that the industry is a terrible one, but are not actually prepared to do anything about it. As long as most people continue to buy budget-priced chicken meat and eggs, there will be no incentive for the producers to change their ways. Why should they? Grumbling about welfare issues but not being prepared to pay more for a better quality product is no use at all.

Organized help

Many countries have established charities dedicated to the rescue and re-homing of ex-battery chickens. These organizations work hard to foster links with commercial poultry farmers so that their "spent" birds can be saved from slaughter and found new homes through a network of regional coordinators.

The organizations charge a nominal fee for each bird, but new owners should be prepared for a shock—newly rescued hens are not a pretty sight, looking more like porcupines than chickens. What feathers the bird

Above *Rescuing hens from the batttery industry is becoming an increasingly popular option.*

has will usually be in a poor state, and combs and wattles will be very pale. This can lead people to think that the birds are unhealthy, although in most cases they are not. They are simply unfit, weak, and in desperate need of some care and attention.

Understandably, it can take ex-battery hens a while to adjust to their new surroundings and freedom. The whole rescue process can be very traumatic for them, and there is always a risk that some may die from the shock of it all. But the majority don't, and their new owners love to see them stretching, flapping, and scratching real earth for the first time in their lives. Being able to take a proper drink is another major event for these birds, which have become accustomed to nipple drinkers that meter out a drop of water at a time. Allowing them access to a bowl of water is another great pleasure.

Taking on birds like this is not for the faint-hearted. Things can go wrong and sometimes birds will die suddenly, which can be upsetting, especially for young family members. But the rewards can be fantastic. The hens recover and re-feather remarkably quickly, and they should start laying again within a matter of weeks. Most will continue to produce a good number of eggs for a further couple of years, all the time basking in a luxurious retirement.

Pests and Diseases

Chickens probably do not suffer from a greater range of diseases than any other livestock, but what can be startling and disconcerting for the novice keeper is the speed with which birds go downhill with problems. When trouble does strike, owners need to be ready to act quickly, seeking veterinary help or the advice of an experienced keeper at the first sign of trouble.

Avoiding problems

Of course, one of the keys to having generally healthy birds is maintaining good levels of husbandry. Looking after your birds well will minimize their stress levels and, thus, the likelihood that they will succumb to diseases in the first place.

By and large, happy birds will be healthy birds, and you can go a long way toward keeping your birds healthy by providing the conditions they want and need: dry, spacious, secure, well-ventilated, and clean housing, good-quality feed, and plenty of fresh water to drink.

Another important factor is close observation. Spending time just watching your birds will not only enable you to learn about their behavior and habits, but it will also help you to spot anything unusual. Important tell-tale signs—birds not eating, being bullied, losing weight, producing diarrhea, for example—should all be noticed quickly by the keen observer, who can then take prompt action to counter the symptoms.

Chickens can suffer from a range of respiratory problems—even a common cold—but diagnosing exactly what the matter is can be difficult, especially for the beginner. The situation is complicated by the fact that many common respiratory diseases share very similar symptoms. Often a vet will be needed to confirm a specific condition. So being aware of the presence of nasal discharges, sneezing, and wheezing—and taking swift action—is another aspect of good husbandry.

Below *This Rhode Island Red illustrates all the key features that a good, healthy chicken should display.*

Inspect eyes for brightness, clarity, and that all-important spark of life.

Damage to the wattles can mean the bird has been fighting.

The tail should be held straight and true. It is important that it points in the right direction; there are a number of faults to watch out for.

Plumage condition should be good, with clean, sleek feathers. Birds should appear active and alert.

Check around the vent for parasite infestation and general dirt.

Spur length is a good indicator of a male bird's age.

The comb should be bright red and free from scabs and discoloration.

Check nostrils for liquid discharge denoting respiratory disease, and watch for beak deformities.

Another favorite place for parasites to gather is under the wing. Also check the shape and condition of flight feathers.

Check for breast bone straightness and the presence of abscesses.

The condition of the legs is important. Scaly leg mite is an infectious nuisance. Watch for "knock knees" and rough scales.

Parasites

Chickens can suffer from both external and internal parasites. Both will cause your birds a good deal of irritation—in some cases even their death—so it is important to keep a watchful eye for their presence and deal with them swiftly and effectively.

Lice

Lice tend to be easier to spot than mites. They will normally be clearly visible crawling around at the base of the feathers, while the eggs will often be spotted clustered on feather stems. You will typically find them congregating among a bird's more downy feathers—under the wings and around the vent are favorite places. Searching for lice infestation should be part of your routine care program. Although they are tiny, the discomfort caused by these irritating creatures can cause death among chicks, so they need to be eradicated at the first opportunity.

Specialized powders are readily available for clearing lice and should be administered exactly as recommended in the instructions. Allowing the birds regular access to a dust bath is nature's way of dealing with the problem, so never underestimate its value. Your chickens should have a box filled with fine, dry soil, sand, or wood ash available to them at all times.

Mites

Mites are a trickier proposition to deal with. There are two main types: the northern fowl mite lives on the birds; the red mite does not. Although they are

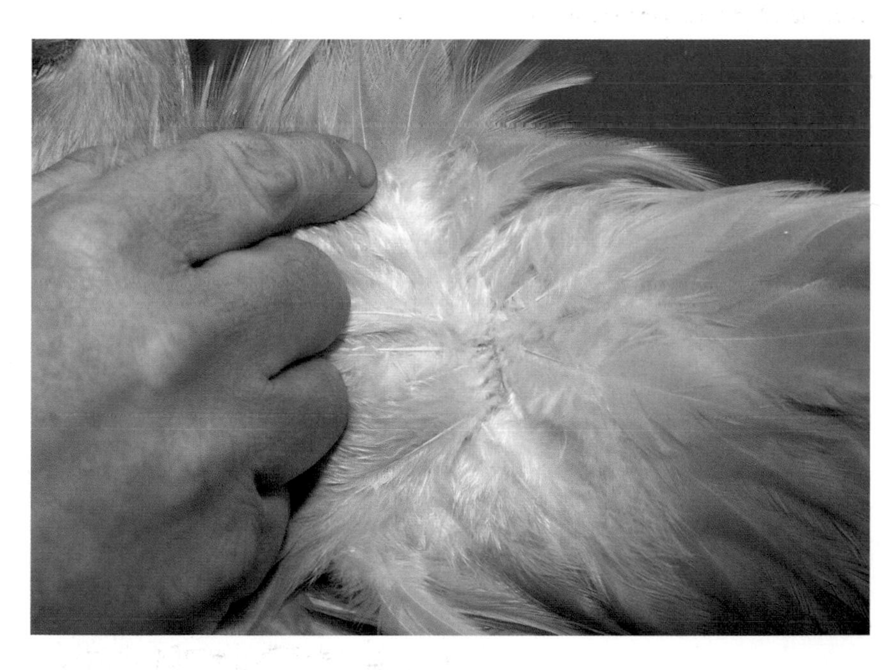

Right *Check your chickens regularly for parasite infestations, and always search right to the base of the feathers.*

parasites that feed on the bird's blood, the tiny, spider-like red mites generally spend their days hiding away in the dark recesses of the chicken coop. They spring into action at night, coming out to feed on the birds as they peacefully roost on their perches. The irritation they cause is significant and can drive birds to the point of plucking out their own feathers. Mites are also perfectly capable of causing a broody hen to desert her nest.

Getting rid of red mites involves more than simply treating the birds: you have got to tackle the root of the problem, which is the henhouse itself. All nooks and crannies will need to be thoroughly cleaned, preferably using a pressure washer, and then disinfected with a top-quality, poultry-safe product. Do not forget to remove and scrub the perches because the undersides are another common hiding place.

Above *Keep roosting areas clean and disinfected. This is a basic requirement of good poultry husbandry.*

Scaly leg

This common condition is just as unpleasant as it sounds. It is caused by tiny mites that burrow into and live under the scales on a chicken's legs, causing them to become inflamed, swollen, and very uncomfortable. If the problem is allowed to develop, the legs will become crusty and often bloody because the scales are gradually lifted away from the skin, essentially by the buildup of excreta from the mites.

Above *Chickens love to use a dust bath, and it is a great way for them to get rid of parasites.*

Like all other parasitic conditions, scaly leg needs to be dealt with as soon as possible. It is extremely contagious, so any birds found to be suffering should ideally be separated for treatment. Consult your vet about the best way of doing this. It needs to be a gentle process, involving the gradual removal of all surface scabbing once it has been softened by soaking in warm, soapy water. After this, a suitable treatment can be applied.

However, never be tempted to lift the scales on a bird's leg. You will not only cause the birds a great deal of pain, but will potentially cause damage. Any scales affected by mited, even after treatment, should remain in place until the bird's next molt, when they will be replaced by new, undamaged scales.

Scaly leg can be promoted by damp conditions. If your birds are found to be suffering, check their house carefully and deal with the problem. Birds with feathered legs are especially susceptible to this common problem.

Worms

Many different types of worm (endoparasites) can infect chickens, and although not all of them lead to problems, some do. Two of note are *Capillaria contorta* and *C. annulata*, which are hair-like worms found in the esophagus and crop. *C. contorta* can also occur in the mouth.

These worms are too small to be easily seen with the naked eye, and the larvae develop inside earthworms before becoming infectious to chickens. They bury their heads in the lining of the mouth, crop, and sometimes the esophagus, causing inflammation and a thick layer of dead and dying cells, which reduces the normal contractions of that part of the gut. This, in turn, results in a loss of appetite and weight.

Tetrameres pose another threat to free-range chickens. Their larvae develop in grasshoppers, beetles, and crustaceans (such as woodlice), and once eaten, they infect the bird's stomach. The female worms are bloodsuckers and can cause anemia. Young birds are most likely to suffer with this particular worm, and no signs are seen in adults.

Treatment for all worms is with regular use of good problem-specific products. Consult your vet for up-to-the-minute advice about what is best to use. Most keepers make a point of worming their birds twice a year.

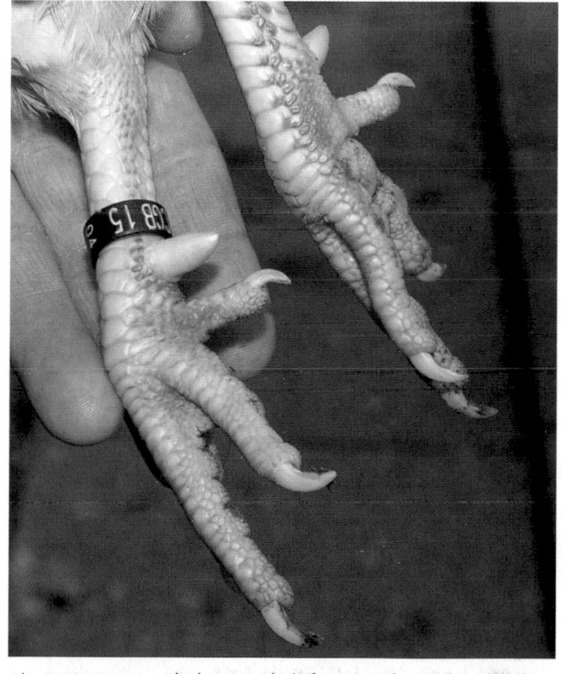

Above *Inspect your chickens regularly for signs of scaly leg. There are no problems here.*

Above *Change the litter in your henhouse regularly so that it never becomes damp and contaminated.*

Diseases and other problems

As with pests, disease prevention is better than cure. Keep your chickens in optimum conditions to prevent problems from arising, and inspect each bird regularly so that you can take appropriate action as soon as you notice anything unusual.

Crop and gizzard impaction

The crop is a large pouch situated at the base of the neck—it is easy to see and feel when it is full. Food is held here while it is softened by the absorption of water, after which it is moved to the stomach by waves of muscular contraction along the esophagus.

A chicken's stomach consists of two parts, and the first of these that the food reaches is called the proventriculus. This is the glandular section, where acid and the digestive enzyme pepsin (used to break down protein) are produced. From here the partially broken-down food is moved into the gizzard, which is very muscular and has strong, rough ridges over its surface to break down the food by mechanical action. The grit that all chickens must eat is used in the gizzard as an abrasive agent to deal with whole grains, seeds, and so on.

The fact that chickens can't chew their food means that they are sensitive to anything that is too long. Long blades of grass and stalks cannot be broken down and might cause problems because they get wound together while in the crop, forming a plug that cannot move on into the esophagus. Then, anything else eaten by the bird gets caught by the blockage until the crop becomes distended and feels hard to the touch. This is crop impaction.

The owner must empty an impacted crop for the bird. The accepted method for doing this involves running some warm water down the chicken's throat (giving it time to breathe), massaging the crop, and then tipping the bird upside down and continuing to massage until the crop contents are expelled through the mouth. Do not tackle this if you are an inexperienced keeper; instead, seek the help of a vet or members of your local poultry club or society, who will be able to teach you the proper technique.

Avoid the problem by keeping your birds on short grass, never feeding them clippings, and keeping them away from areas of uncut ground with tough, fibrous grasses. If a large number of birds are having this

Above *Crop impaction can be felt for by hand.*

problem, a mixture of Epsom salts, molasses, and vegetable oil can help.

If the crop keeps refilling, even if the bird is on short, soft grass or indoors, the problem may well be a gizzard impaction. Here, the grass or other foreign matter that has gotten through the crop can form a "rope" that blocks the top of the intestinal tract. This is a much more serious condition than crop impaction and one that can often prove fatal.

Although birds of any age can be affected, young chickens out on their first grass are especially susceptible because their gizzards may not have developed sufficiently to cope with grass. The problem can also occur with birds that are on grass but not getting enough dietary grit.

Above *Take care with the sort of grass you feed, particularly to young ones.*

Coccidiosis

This is a potentially fatal disease of young chickens between three and six weeks old, but it can also affect adults. Typical signs include depression, huddling, fluffing-up of the feathers, white soiling of the vent, diarrhea, and paleness.

All hens will possess some coccidia—a low level of infection is common. It is caused by a protozoan (a single-cell parasite) and is spread among the flock via feces. It can survive out of a bird (inside a henhouse) for years. It is resistant to many disinfectants and is easily transported between groups of birds on clothing, boots, feed buckets, insects, or other animals.

Factors that increase the likelihood of coccidiosis occurring include overcrowding, keeping young chicks where older birds have been, and rearing birds in a warm, wet, and under-ventilated environment.

Chicks have no immunity from the hen and only gradually acquire resistance after a low-level exposure, going on to become fully immune at about seven weeks old. However, immunity will be to only one strain of coccidia, so different species will cause infection.

To minimize the risk of infection for the first 16 weeks, you can feed a chick crumb containing an anti-coccidial agent. Of course, good hygiene and management methods will still be necessary. There is also a vaccine for chicks up to two weeks old, but this is only practical for birds reared on a commercial scale.

Marek's disease

This is one of the most common diseases of poultry, affecting both chickens and turkeys. It is controlled commercially with vaccination at day-old (chicks up to 48 hours old), but it still causes problems—and fatalities—among birds kept on a domestic basis. Incidentally, Sebrights and Silkies are the two most susceptible pure breeds.

It is caused by a herpes virus and is spread by feather dander—that is, microscopic particles that can travel miles on the wind. The virus matures and becomes infectious in the feather follicles and is then released with the feather dander. Once in the environment, it will remain infectious for at least a year, and carrier birds will shed Marek's virus with no signs of infection themselves.

The disease causes the growth of tumors in the nervous system, resulting in various degrees of paralysis—legs, wings, and head are typically affected. Death is usually due to suffocation (in large flocks) or dehydration because the bird is unable to drink.

Not all infected birds die, but those that survive will be carriers and may succumb in the future if they become stressed. Pullets are more susceptible than cockerels. As chicks get older they develop age resistance, so that after about one month they have a degree of natural immunity. Adult birds that are carriers of Marek's disease can be identified by looking closely at their eyes: if the color of the pupil flows into the iris, the bird is likely to be a carrier.

There are two main ways of dealing with Marek's: vaccination at day-old and breeding for resistance. For the domestic keeper, however, vaccines are expensive, difficult to handle, and wasteful. The best approach is to reduce the level of viral challenge by sweeping or hoovering (wetting first) the poultry house daily to remove as much of the dander as possible and to reduce the challenge to the birds.

The susceptibility to Marek's varies between breeds and lines of birds. A breeder's aim should always be to breed the fittest, healthiest stock possible and to consider the finer points of the standard after that. Of course, most of the points breeders aim for are those that normally take the bird further away from its original ancestor, the Jungle Fowl. Breeding for recessive qualities—large combs, silky feathers, daily eggs, and so on—automatically brings with it other, less desirable recessives, including reduced resistance to disease.

Brought-in birds should be assumed to be carriers, but quarantine alone will not work because they have to be integrated into the flock at some point. Good management is vital in order to keep down the stress levels, which helps to minimize the viral load.

Marek's disease is a serious problem that all keepers should be aware of.

Mycoplasma

This is a serious respiratory disease, and there are a number of forms that can strike chickens. Many are present naturally in the birds, becoming apparent only when something else, such as a virus, weakens the bird. There are other types that are pathogenic and affect otherwise healthy birds. Even these, however, normally require some other predisposing factor, such as poor air quality, poor hygiene, or high levels of stress, before taking hold.

The type that causes the most problems for chickens, *Mycoplasma gallisepticum*, is spread by egg transmission, aerosol infection (from infected droplets of moisture from sneezing or coughing birds), and direct contact with infected and carrier birds. It causes chronic respiratory disease and air sac syndrome. Birds affected will sneeze, cough, and produce some nasal discharge. In the worst cases—air sac syndrome—the air sacs fill with matter, the birds become droopy and lethargic, show a reduced appetite, and lose weight rapidly.

Above *Cyanosis—blueing of the comb—is associated with avian influenza.*

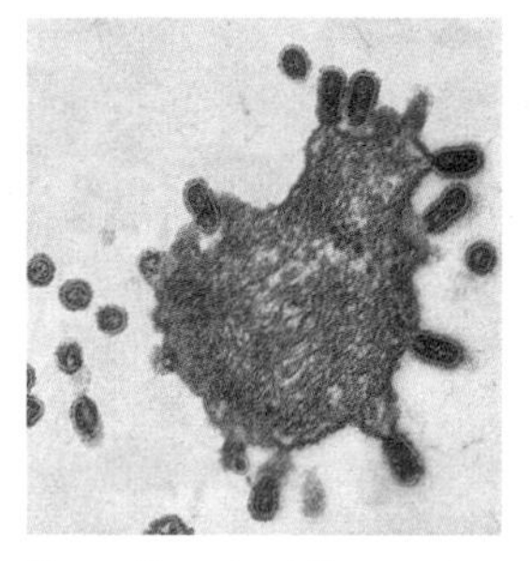

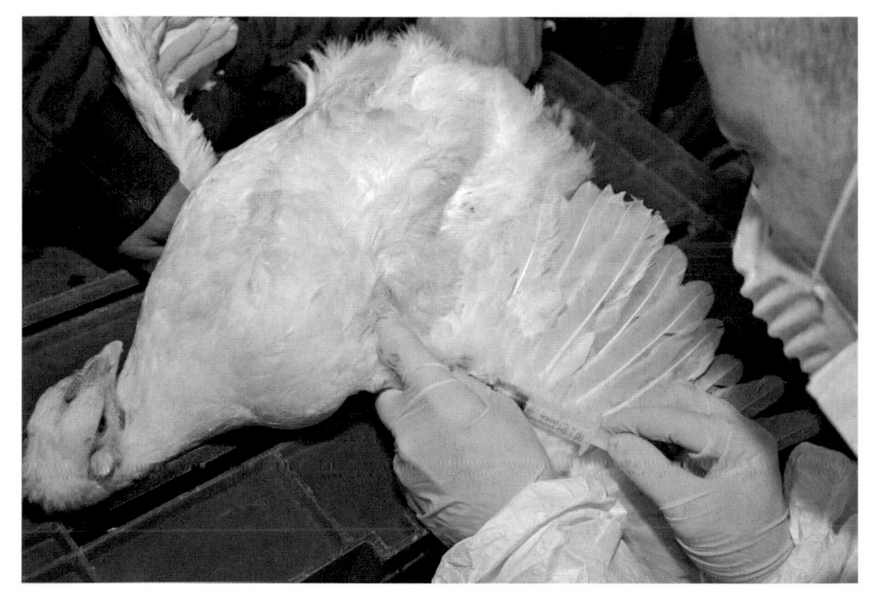

Above *The avian influenza virus, H5N1—deadly for chickens and potentially humans as well.*
Right *Vaccination against avian influenza has yet to be proven 100 percent effective.*

Of the antibiotics licensed for use in poultry, tylosin is the most often prescribed, although resistance to it is increasing. Dipping or injecting the eggs can reduce the number of positive chicks hatched according to some research, but this method is not always reliable. Consult your vet immediately if you suspect that any of your chickens is infected.

Poultry sales and auctions held in markets are a true mixing pot of infection—the birds are stressed, mixed with all sorts of other species, and then put into a new environment. If you must buy birds from one of these sources, quarantine them, give them constant fresh air, clean water, adequate ambient temperature, and comfortable, clean, and dry bedding.

Avian influenza

This is a normally fatal disease for chickens. There are many strains, but most recently the virulent H5N1 type has been causing severe problems in the Far East and parts of mainland Europe, where millions of birds have been slaughtered in a bid to control the outbreak.

The avian influenza virus is highly contagious and is caught following contact with feces or expelled air from infected birds. It is spread by all birds—migratory waterfowl are particularly effective carriers—which makes it extremely difficult to contain. The practical signs of infection are essentially limited to a "blueing" of the bird's comb and wattles and sudden death.

The only way to prevent chickens from catching avian influenza is to stop them coming into contact with infected birds. In the case of domestic poultry, this means keeping them isolated from all other birds, using enclosed runs with wildbird-proof netting, and (ideally) solid roofs, but for many keepers this simply is not a practical proposition for reasons of both space and cost.

The contagious nature of the disease and the risk that it will mutate into a potentially deadly human form mean that the authorities are bound to take outbreaks extremely seriously. Exclusion zones will be set up around the center of outbreaks, and the birds within are all likely to be culled.

Glossary

air space The air space at the broad end of an egg, which indicates freshness and, during incubation, humidity levels.
ark An A-frame, portable poultry house. Such henhouses may have slatted floors or rest directly on the ground.
auto-sexing The method by which a chick's sex may be distinguished soon after hatching by the color of its down.
axial feather The feather that lies between the primaries and secondaries on a chicken's wing.

band A stripe of color (often black) that runs across a feather.
bantam A miniature fowl, originally one-fifth the weight of the large equivalent breed but now one-quarter of the weight. Some bantam breeds, known as true bantams, are naturally occurring and have no large fowl equivalent.
barring A pattern of equally sized, alternating stripes of two distinct colors running across the feather.
bay A reddish-brown plumage color.
beard The feathers found around the throat of chickens, such as Faverolles.
beetle brows Pronounced, distinct eyebrows as on the Malay.
booted Feathered shanks and toes, as on the Brahma and Cochin.
brassiness A yellowish tinge to the feathering.
broiler A young chicken specially bred for the table.
brooder An externally powered heater used for the artificial rearing of chicks.
broody or **broody hen** A hen that has stopped laying in order to sit and naturally incubate a clutch of eggs, which may be her own or those of another hen.
bumblefoot A swelling of the foot, commonly caused by unnecessary jumping, such as from high perches.

candling A method of determining if a hatching egg is developing. The egg is held in front of a strong light.
cape The feathers between the neck and shoulders.
carriage The general appearance and visual impression given by a chicken.
caudal Relating to the tail. The parson's nose, the fatty, fleshy protuberance at the tail-end of a fowl, is correctly called the caudal appendage.
chalaza (pl. chalazae) One of the two coils of albumen that support the yolk, keeping it central inside the egg.
cock A male bird after his first molt.
cockerel A male bird from the current year's breeding—i.e., less than 12 months old.
comb The fleshy protuberance on top of a chicken's head. It takes various forms, including single, rose, cup, leaf, pea, and V-shaped.
coverts The small feathers on a chicken's wings and tail that surround the bases of the larger feathers.
crest The arrangement of feathers on top of the head.
crop The food collection sac inside the base of the neck, where food is softened before passing on to the ***gizzard*** and through the digestive system.
crossbreeding The mating of different breeds or varieties.
cuckoo banding or **cuckoo barring** Uneven and irregular plumage marking, where two colors overlap indistinctly, as on Cuckoo Marans.
culling The humane killing of ill or unwanted birds.
cushion The raised mass of soft feathers on a hen's back, at the base of the tail. *See also* ***saddle***.

day-old A chick up to 48 hours old.
dead in shell A chick that has died before hatching.
debeak To remove the tip of the upper beak, a practice carried out by commercial rearers to prevent hen pecking.
double-laced Two narrow, parallel lines of contrasting color around the edge of a feather.
down The soft, fine covering on a young chick. The fluffy layer below the main feathers on an adult bird.
dual purpose A bird that is bred to be both an egg producer and a table fowl.
dub To trim or cut off combs, wattles, and earlobes to leave a smooth head.
dust bath A container of clean, dry sand, fine, dry earth, or wood ash in which a chicken is able to "bathe." Such activity is vital and helps chickens to remove external parasites from their bodies.

earlobes Folds of skin that hang below a bird's ears; they vary in shape, size, and color.
egg tooth The tip of a chick's beak, which initially pierces the shell at hatching time. It falls off after use.

flight coverts The small feathers that cover the base of the wing primaries.
flight feathers *see* ***primaries***.
fluff The very soft feathers typically found around the thighs on breeds such as the Cochin.
free range Traditionally referring to a stocking density of no more than 50 birds to 1 acre (0.4 hectares). It is these days a much-abused term.
frizzled Feathers that curl strongly, often to face toward the bird's head. The Frizzle is a bantam breed.

gizzard The muscular compartment within a chicken's digestive system that contains the grit that is used to break down the food.
ground color The background color on a patterned bird.
gullet The esophagus or tube between the throat and the ***crop***.
gypsy-face The traditional name for a dark purple or mulberry-colored face.

hackles The long, narrow feathers typically found on a bird's neck and shoulders. The pointed saddle feathers on a male.
harden off To wean young birds away from an artificial heat source or broody hen.
hard feather The tight, smooth feathering, with no layer of down, that closely follows the contors of the bird's body. Game birds tend to be hard feathered. *See also* ***soft feather***.
heavy breed A breed in which the mature female weighs more than 5½ lb (2.48 kg). Such birds frequently have Far Eastern origins. *See also* ***light breed***.
hen The term for a female chicken following her first adult molt.
hen-feathered A description of a male bird that does not have sickle feathers in the tail or pointed hackle feathers elsewhere.
hind toe The rearward-pointing toe on a chicken's foot.
hock The joint at the junction between the thigh and the shank; it is equivalent to the human knee.
horn comb A two-pronged comb, often referred to as a V-shaped comb.
hybrid A commercial cross-breed created for egg or meat production.

in-breeding The crossing of closely related stock.

keel The blade of a chicken's breastbone.

lacing The contrasting edge color to a feather.
leader A backward-pointing spike found at the back of a ***rose comb***.
light breed The opposite of ***heavy breed***; a breed that is generally flighty, quick to feather, and light in weight.
litter The floor covering of a chicken henhouse and run; several materials are suitable.

mandibles The upper and lower parts of the beak.
mash A mixture of ground poultry food, which can be fed wet or dry, warm or cold.
meat spot A small, often reddish mass found in an egg. It is caused by pieces of tissue coming away from the oviduct and passing into the egg. A meat spot that is unsightly but harmless.
mottled Feathers with tips or spots of a contrasting color.
molt The annual shedding of feathers, during which hens stop laying. It can taken up to three months for all feathers to be replaced.
muff Feather tufts on either side of the face, usually attached to the beard.
muffling A general term for all face feathering, apart from the crest on top of the head.

pellet A small, shaped piece of food that contains a balanced range of ingredients; it can be tailored for different purposes, including breeders, layers, fattening, and finishing.
penciling The fine markings or stripes across a feather.
pin feathers The first feathers to come through after the annual ***molt***.
point-of-lay (POL) The term used to describe hens, usually about 18 weeks old, which are about to lay their first eggs.
primaries The ten main flight feathers on a wing; they are not visible when the wing is folded.
pullet A female chicken from the current season's breeding—i.e., less than 12 months old.

quill The hollow stem that links a feather to the body.

roach back An undesirable deformity that shows as a humped back.
rose comb A broad and almost flat-topped comb that is covered with raised nodules (known as work). The ***leader*** extends backward from the comb.

saddle The rear portion of a male bird's back; the equivalent of the hen's ***cushion***.
scales Small, thin, overlapping flakes that cover the legs and feet of a chicken and are replaced during each molt.
secondaries The inner group of shorter quill feathers in the wings, which are visible even when the wings are folded. *See also* ***primaries***.
self-color A uniform color that is unmixed with any other color.
serration A "sawtooth" indentation on a single comb.
shaft The ***quill*** or stem of a feather.
shank The part of the leg between the ***hock*** and foot.
shoulder The upper part of the wing, nearest the neck.
sickle feathers The long, curved feathers in a male bird's tail. The name often refers only to the top two feathers.
side sprig An extra spike growing from the side of a ***single comb***.
single comb A narrow, spiked and usually upright comb; the color, size, and number of serrations vary from breed to breed.
slipped wing A condition in which the primary feathers hang below the secondaries when the wing is folded.
soft feather Loose, fluffy plumage. *See also* ***hard feather***.
spangling The marking produced by a spot of contrasting color at the end of each feather.
splashed feather A feather irregularly marked with a contrasting color.
split crest A divided crest that falls over on both sides.
split tail A tail that has an obvious gap in the center at the base.
sport A bird with a naturally occurring characteristic, such as color, that differs from the species. A sport may be the result of genetic mutation.
spur The horny growth on the rear of the leg, above the foot, of male birds and some females.
squirrel tail A tail that curves forward and comes close to touching the back of the chicken's head.
strain A single family of birds from any breed or variety that has been carefully bred over a number of generations.

striping The markings down the center of hackle feathers.
stub A short, partly grown feather.
sword-feathered A bird with sickle feathers that are only slightly curved.

tertiaries The wing feathers closest to the bird's body.
thigh The section of leg above the ***shank***; it is covered with feathers.
toe-punching A method of identifying chicks by punching holes in the web between the toes.
trio A group of one male and two females birds, often bought together for breeding purposes.
type How a bird looks; his characteristic shape.

undercolor The color seen when a bird is handled and the outer feathers are lifted.
utility A breed that is recognized for its ability to lay eggs or provide meat.

variety A distinct branch of a breed that is set apart by an obvious marking or color difference.
vent The rear orifice through which eggs and bodily excretions are passed.
vulture hocks Stiff, angled feathers that grow from the hock joint and point down and backward.

wattles The fleshy appendages that hang as a pair from below the beak on each side of the face. They are larger on male birds.
web The flat part of a feather. The skin between the toes.
whiskers The feathers that grow from either side of the face.
wing bar A visible line of a darker color across the middle of a wing.
wing bay The triangular section of the folded wing, between the ***wing bar*** and the point.
wing bow The upper or shoulder part of the wing.
wing butt The end of the primaries; the corners or ends of the wing.
wing coverts The feathers that cover the roots of the secondary quills.
wry tail A tail that does not point directly backward but veers to one side or the other.

Index

Acknowledgments

T.F.H. Publications

President/CEO Glen S. Axelrod

Executive Vice President Mark E. Johnson

Publisher Christopher T. Reggio

Production Manager Kathy Bontz

US Editor Mary E. Grangeia

T.F.H. Publications, Inc.
One TFH Plaza
Third and Union Avenues
Neptune City, NJ 07753

First published in Great Britain in 2006
by Hamlyn, a division of
Octopus Publishing Group Ltd

Executive Editor Sarah Ford

Senior Editor Jessica Cowie

Deputy Art Editor Geoff Fennell

Designer Janis Utton

Senior Production Controller Martin Croshaw

Picture Researcher Sophie Delpech

Photo Acknowledgments

Main photography © Chris Graham.

Other Photography:
Alamy/Simon Belcher 199 right; /Che Garman 190–191, 194 bottom; /Tim Graham 47 bottom; /Sally and Richard Greenhill 195 right; /D. Hurst 16; /Keiji Iwai 72; /Kim Karpeles 59 top; /Renee Morris 32 left; /Jack Sparticus 29.
Corbis UK Ltd/Robert Dowling 136, 165; /Lars Langemeier/A.B./Zefa 66; /James Marshall 108; /Robert Pickett 64; /Herbert Spichtinger 54-55.
Frank Lane Picture Agency/Jim Brandenburg 32 right; /Nigel Cattlin 47 top; /David Hosking 28; /Frank W. Lane 157; /Gordon Roberts 188; /Terry Whittaker 31.
Masterfile/Gary Rhijnsburger 70-71.
N.H.P.A./Joe Blossom 33, 137.
Science Photo Library/Eye of Science 199 left.
John Tarren/David Scrivener Archive 170, 171, 176, 177, 178, 179.
Warren Photographic 130.

Special thanks to Tony Beardsmore, Terry Beebe, Ken Blow, Colin Clark, Richard Cowans, Alan Fearn, Rachel Graham, Nancie Hutchinson, Andy Marshall, Stuart May, Darren Peachey, Geoff Silcock, Hans Schippers and Middle Farm, Firle, East Sussex.